CONCILIUM

concilium 1995/5

ECOLOGY AND POVERTY

Cry of the Earth, Cry of the Poor

Edited by

Leonardo Boff and
Virgil Elizondo

SCM Press · London
Orbis Books · Maryknoll

Published by SCM Press Ltd, 9–17 St Albans Place, London N1
and by Orbis Books, Maryknoll, NY 10545

ISBN: 0 334 03034 X (UK)
ISBN: 0 88344 886 6 (USA)

Typeset at The Spartan Press Ltd, Lymington, Hants
Printed by Mackays of Chatham, Kent

Concilium Published February, April, June, August, October, December.

Contents

Yves Congar *In memoriam*
JEAN-PIERRE JOSSUA vii

Editorial Ecology and Poverty: Cry of the Earth, Cry of the Poor
LEONARDO BOFF and VIRGIL ELIZONDO ix

I · Some Problems 1

The Present Socio-Economic System as a Cause of Ecological Imbalance and Poverty
JULIO DE SANTA ANA 3

Ecological Consciousness in Amazonia: The Indigenous Experience
BERTA G. RIBEIRO 12

Sacred Earth: Mesoamerican Perspectives
SYLVIA MARCOS 27

II · A Theological Reflection 39

The Cry of the Earth? Biblical Perspectives on Ecology and Violence
CHRISTOPH UEHLINGER 41

Spirituality of the Earth
JULIA ESQUIVEL VELÁSQUEZ 58

Liberation Theology and Ecology: Alternative, Confrontation or Complementarity?
LEONARDO BOFF 67

Some Premises for an Eco-Social Theology of Liberation
JOSÉ RAMOS REGIDOR 78

III · New Horizons 95

Towards a New Paradigm of Production: From an Economy of Unlimited Growth Towards One of Human Sufficiency
BASTIAAN WIELENGA 97

Ecology from the Viewpoint of the Poor
EDUARDO GUDYNAS 106

Principles for a Socio-Environmental Ethic: The Relationship Between Earth and Life
CHARLES RICHARD HENSMAN 115

The Theological Debate on Ecology
ROSINO GIBELLINI 125

Special Column 135

Contributors 141

Yves Congar

In Memoriam

Our master and friend died in the night of 21/22 June 1995 at the age of ninety-one, after a long illness which in the last stages had become intolerable. Our sadness at losing him is mixed with the gratitude that we feel at having known him, at having received so much from him, at having been able to see all that he brought to the church.

He was one of the founders of *Concilium* and had great hopes for this journal. His work on the Foundation, his articles, his faithful reading of the issues, his reflection on the plans for future numbers, his contribution at the general assemblies, made him one of the key figures. He followed confidently the series of developments which *Concilium* has undergone – though he could show a degree of irritation at some attitudes or expressions – and he recently asked me again about the plans for reform which were to be adopted during this year.

It is impossible here to sum up his career and his writings, but in any case they will be well known to readers. I shall recall just one facet of them. From 1930, his love of the church, his desire to serve it by ecclesiological thought and his struggle for reform took the form of one essential concern, that of the unity of Christians. Granted, Yves Congar had to retire from this area, and he devoted himself to many other works – which coincided with the preoccupations of the Second Vatican Council, hence the major role he was able to play in it – but this was always with the aim of preparing for unity. In this sphere his thought never ceased to progress, from *Chrétiens désunis* in 1937, through *Chrétiens en dialogue* in 1964, to *Diversity and Communion* in 1982, which bears witness to a marked advance. However, such a passion for the church and its unity was not ecclesiocentric. It was aimed at proclaiming and receiving the gospel. It took account of a world which he looked on in an increasingly open way, accepting its consistency and its autonomy.

These views and these concerns remain ours, in the diversity of our cultural situations, and with our acknowledgment of this goes a commitment to continue his efforts.

Jean-Pierre Jossua

Editorial

Ecology and Poverty: Cry of the Earth, Cry of the Poor

From its beginnings, ecological reflection has developed on several lines. It has moved beyond the *conservationism* that was reduced to conservation of endangered species, and the *preservationism* that led to the creation of ecological reserves with rich biodiversity, interesting for ecological tourism, but limiting ecological concern to these areas only, while we go on plundering nature outside them. It has produced a critique of *environmentalism*, which restricted ecology simply to the natural environment, forgetting human beings, who are part of this environment. It has pointed out the limitations of *human ecology*, on the grounds that it sees only the relationship of the individual human man/woman with nature without taking account of the fact that this individual is always particular and social and establishes historico-social relationships with his/her surroundings. It then issued in a *social ecology*, which studies socio-cultural systems in interaction with environmental systems, considering forms of exploitation of or benevolence to nature, especially in relation to the social mechanisms that produce rich and poor, participants and excluded, and so on. Finally, a *mental* or *deep ecology* has developed, which stresses the importance of mental/cultural structures (ideas, values, neuroses, prejudices, etc.) to aggressive or respectful relationships with nature and with human beings within the natural system.

The quest today is increasingly for an *integral ecology* that can articulate all these aspects with a view to founding a new alliance between societies and nature, which will result in the conservation of the patrimony of the earth, socio-cosmic well-being, and the maintenance of conditions that will allow evolution to continue on the course it has now been following for some fifteen thousand million years.

For an integral ecology, society and culture also belong to the ecological complex. Ecology is, then, the relationship that all bodies, animate and inanimate, natural and cultural, establish and maintain among themselves and with their surroundings. In this holistic perspective, economic, political, social, military, educational, urban, agricultural and other questions are all subject to ecological consideration. The basic question in ecology is this: to what extent do this or that science, technology, institutional or personal activity, ideology or religion help either to support or to fracture the dynamic equilibrium that exists in the overall ecosystem?

On the subject that concerns us – ecology and poverty – it is important to stress the contribution made by social ecology, since this deals directly with these questions. At the first United Nations international conference on the environment, held in Stockholm back in 1972, and more openly at the Rio Earth Summit of 1992, which first conceived and then developed the concept of 'sustainable development', two basic visions emerged.

The industrialized nations of the North tend to put forward, though not exclusively, an environmentalist outlook, with the focus on the environment, and giving insufficient consideration to human beings. The poor countries of the South generally put forward a political and social view, putting human beings at the forefront, though in interaction with the environment, paying special attention to the poor of the earth, who are the victims of ecological aggression. For these, social deprivation, the 'belts of misery' around the great cities of the world, disease, lack of housing and education, the concentration of land in a few huge ranches, the technology of agrobusiness, the international trade in foodstuffs, the greenhouse effect, the hole in the ozone layer, the pollution of air and water and the threat to biodiversity all constitute vital questions on the ecological agenda.

We are faced with two types of injustice: social injustice, the result of attacks on workers' rights, and those of the unemployed and other forms of underclass; and environmental injustice, which is violence against the environment, against the atmosphere, against the ozone layer and against the seas and rivers. Social injustice affects individuals and social groupings directly. Environmental injustice affects them indirectly, but with immediately harmful consequences, since the decline in the quality of their surroundings produces social tensions, violence, disease, malnutrition and even death. It is not only the biosphere that is affected, but the global ecosystem of the planet Earth. A minimum of environmental justice is essential for a minimal social justice as well as for the preservation of the *dignitas terrae*.

Today, nature's most threatened creatures are not the whales or the giant pandas of China, but the poor of the world, condemned to die of hunger and disease before their time. Not everyone is conscious of this perversity.

In the course of their historical and cultural development, human societies have always interfered with the environment. There was aggression, but also a sort of pact of respect and collaboration. Human ingenuity produced improved species used in diet, as with potatoes, maize and tomatoes – to take only products of the Aztec, Maya and Inca cultures of America. Ecological disturbances were not very serious. In the last four centuries, however, with the coming of the industrial era, aggression against nature has been carried out systematically and pitilessly.

The whole of the earth has been reduced to natural capital, an accumulation of resources for growth and profit, first for those who have private ownership of these resources and then for the rest. And workers have been reduced to human capital. The present-day result is devastating. World-wide, social relationships destroy nature and exclude people. An unjust and humiliating attitude to the earth prevails. The earth in turn can no longer support the mechanisms of destruction and death implanted in it. Either we change course, or we are heading straight for great ecological calamities. The earth is crying out and the poor are crying out, both victims of both social and environmental injustice.

The existence of rich and poor in our societies is in itself a form of ecological aggression. The rich consume too much, wastefully and without thought for the present or future generations; they have set up a technology of death to defend their privileged position, with nuclear and chemical arsenals that could, at worst, bring about biocide, ecocide and even geocide; furthermore, they defend a production system whose inner logic makes it a predator of nature. The poor, victims of the rich, consume less and, in order to survive, live in unhealthy conditions, cut down forests, contaminate waters and soil, kill rare animals and so on. With greater social justice they would be able to operate better environmental justice.

How can we obtain a socio-economic system that will produce a decent sufficiency for all, within a development model worked out with nature and not against it, and in which the idea of the common good will also involve the common environmental good, that of the air, seas and rivers, living beings, the whole environmental landscape? This is the great challenge raised by the cry of the poor and the cry of the earth.

To tackle this important theme, we have invited contributors from all over the world. It was vital to hear the witness of primary cultures, since they reveal an attitude of great reverence for nature, which is also reflected in their productive processes. The most archaic can become the most modern, in that it pays urgent attention to a universal demand. These reflections equally seek to reinforce the conviction that, in a healthily environmental perspective, politics must be subject to ethics and ethics to a mysticism or spirituality – meaning, to a basic experience of re-binding with everything and with the Thread that binds everything upwards, God.

Leonard Boff
Virgil Elizondo

Translated by Paul Burns

I · Some Problems

The Present Socio-Economic System as a Cause of Ecological Imbalance and Poverty

Julio de Santa Ana

At the end of the twentieth century we have a world of surprising contrasts: a minority of the planet's population has great power to accumulate wealth, enabling them to enjoy an excessive material affluence, whereas most of the inhabitants of the globe have to struggle to survive, in conditions of miserable poverty. Despite the enormous increase in the gross world product over the last fifty years (since the war of 1939–45), approximately 20–25% of the earth's inhabitants live below the poverty line. There are vast areas of the world where the majority exhaust themselves merely in the struggle to stay alive. The price they pay is too high. It is also a scandal when it is compared with the affluence of the well off.

The problem cannot be reduced to a contradiction between rich and poor countries. Although the injustice of the situation can be seen more clearly among inhabitants of the South, we must recognize that it also exists in rich countries of Western Europe and North America, where the number of unemployed is increasing. It is a phenomenon which mainly affects the young, particularly those of African, Asian, Latin American and Caribbean origin and including native peoples (like the 'Native Americans' in the USA). This problem of the 'workless' cannot be explained by the economic recession. In fact the recession that existed at the beginning of this decade is almost over. Despite the existing financial unrest (which was plainly in evidence with the Mexican crisis and its consequences, with scandals in prestigious banks, such as Barings in Britain and Crédit Lyonnais in France) there has been reasonable economic growth. Nevertheless there is no sign of unemployment decreasing. That means there is production of wealth from which some are excluded.

This has led to denunciations of 'the irrationality of the rationality'[1] of the current economic system. The same phenomenon appears when we look at environmental deterioration: pollution, growth in conditions threatening to health, decrease in the protective ozone layer around the poles of the planet, acid rain which causes the death of important plant life are some of the phenomena showing the senseless way in which the environment is being exploited. There is an absurd insistence on reducing economic processes simply to profit-making.

The UN Summit on the Environment and Development in Rio de Janeiro, Brazil, in June 1992, and the Summit on 'Social Development' that has just taken place in Copenhagen, Denmark in March 1995, were both occasions to look at these surprising contrasts. Side by side with these summits we saw a growing denunciation of 'the irrationality of the rationality' of the current system, and also an increase in non-governmental bodies trying radically to correct these senseless tendencies in the system.

The present system and its tensions

The system we are living under is the result of historical changes, which took place through processes of 'long duration'.[2] This system is the result of the commercial, political and cultural expansion of the West over the last four or five centuries. Before this system arose and was formulated, the greatest power on the planet was concentrated in the Far East. However, neither China nor Japan became 'world' powers. Their influence was felt over the vast areas of the Pacific, attracting some Arab and Western merchants.

The situation began to change with the expansion of the European West, and, as mentioned above, had three interrelated dimensions: commercial, political and cultural. It is a process with an inbuilt drive to dominate. It had a 'Faustian' spirit, recognizing no limits, which cultivated the market in order to obtain advantages. When people acting in this spirit arrived in a land where other peoples were already living, they tried to impose themselves upon them, conquering them by what they called cunning and strength. This conquest often imposed new names on the geographical locations conquered: the original indigenous names were forgotten and were replaced by Western names. This practice of domination, conquest and colonization enabled countries which until the sixteenth century had been poor and backward to begin a process of accumulating wealth, which in turn gave them the opportunity to make great advances in knowledge and technology. Western science was in accord with the guiding spirit of domination. This science reflects the exploitative nature of those who are conquerors and colonizers.

In our day this system is seen as a global entity. Those who control it and profit from the benefits of its processes of production, circulation of a wide variety of goods, trade and finance, are not only people in the West. The 'ruling class'[3] is made up of persons and economic groups in many countries. Therefore in our time the dominant systemic culture is cosmopolitan. It does not express local or national interests, but an attempt to control the world. This is what enables it to present itself as a 'universal culture'. In reality, it is a culture trying to dominate and control everything, without limits, in accorance with the 'Faustian' spirit driving it.

This world system has a certain shape: it has a centre to which most of the accumulation gravitates. This is where wealth and power are concentrated. Around this centre there are 'inner circles' which also manage to accumulate wealth by supporting the central power. Far off on the 'outer circles' there are the conquered, those who are economically, politically or culturally subordinate. Apart from the exceptions who manage to maintain beneficial links with the centre and its 'inner circles', those on the outer circles work in the service of the system's established power. Between the centre and the 'inner circles' on the one hand, and the outer circles on the other, there are webs with points of intersection linking the former to the latter. At these points some accumulation of wealth also occurs. This was the case with the great 'port cities', which nowadays appear to have been replaced by 'free zones', financial havens where international economic and fiscal power can be operated in relative freedom, i.e. with impunity.

Advances in information technology have helped to consolidate the system. First, it presents itself as a global 'thing'. Secondly, its clearest expression is 'the market', that is, the sum of agreements, contracts, accords and conclusions through which human beings exchange their goods. 'The market' is an artefact created by individuals and their institutions. Logically, those who have the most power are the ones able to determine the shape of the 'market' in accordance with their interests. In the history of the system it is a fact that the central powers have profited from these advantages. This was the case with the Western colonial empires (principally the Netherlands first of all, and then Britain) and the USA. Thirdly, in the present period, when the USA is the most indebted country in the world, the centre of the system is tending to shift: it may move towards the Far East, where we are seeing great economic dynamism (in which case it could centre on Japan, but with 'inner circles' which would include Korea, Taiwan, Hong Kong, China and other South East Asian countries), or towards Western Europe (whose major power zone lies between London, Paris, Berlin and Milan). This inter-systemic transition is one of the most important causes of the present instability. And here I should point out that this instability has the greatest impact upon the outer circles and the points linking it with the centre.[4]

I have already referred to the 'irrationality of the system's rationality'. The system is irrational, unbalanced, shaken by various types of disorder. We can see this more clearly when we analyse it from the outside, and especially when we interpret an analysis with criteria which are not priorities in the system's own logic, for example, when we take into account the need for equality and fairness as imperatives for the existence of freedom and justice. Then we see that the system is neither fair nor just. Moreover, when the system speaks of freedom, it does not mean freedom for most human beings but, principally, freedom for 'the market'. That means the freedom of those who organize the system's processes by means of the power that they hold in order to satisfy their own interests. The system is unjust because it practises *exclusion*. Those who cannot pay the price necessary to enter the system remain outside. The different groups of those excluded (for example, most of the peoples of Africa between the Sahara and the Zambesi, most of the peoples of Central America, north-east Brazil, the South Pacific islands, north-east India and Bangladesh, and among these peoples, above all the women) are victims of the system's logic.

The force of the 'principle of exclusion', belonging to the logic of the system, is felt with terrible violence. It causes suffering. It leads to a perversion of life for the poorest, who in order to survive often have to do things which are against their own dignity. This violence can be felt in many areas of social life: not only in international, inter-ethnic or inter-cultural relations, but also in its anonymous manifestations in the great cities, both of the North and the South.

Despite the fact that the force of the system's power is very great, there are social sectors which realize what the situation is and try to do something about it. That is, just as there are powerful systemic agencies to try (consciously or unconsciously) to impose the 'spirit' of the system – ruthless and often disloyal competition, individualism, as opposed to solidarity and altruism – there are anti-systemic movements seeking to make socio-economic processes work in a different way. They answer exclusion with the need for sharing and inclusion. In response to competition, which is often closely linked to resentment, they point out the need to recognize that life closes in upon itself when it practises 'social

Darwinism'. They understand that the socio-economic reality can become favourable to the confirmation of life when we recognize other men and women and give them our attention, respect and love.

Of course things are not as simple as this. The anti-systemic movements do not escape the system. It may even happen that they are co-opted by the system. This has sometimes happened, and it can happen to any movement fighting for social justice, or for justice in gender or race relations, or for 'national liberation', or for the defence and promotion of human rights, or for greater respect for the environment. We have to realize that in real life things are not as clear as they seem in theory. There is an area of ambiguity of which we must become aware. Above all this will enable us to achieve greater coherence between convictions and intentions, between ethos and actions.

Despite this possibility of sliding into ambiguity, the fact remains that these anti-systemic movements do introduce tensions and disorders into the system, which in some way force it to consider having to check its excesses. Even more important, they prevent the system becoming closed in upon itself and totalitarian, which may occur if its own dominant tendencies are carried to excess.

The sacrifice of the poor

I have already said that this system is characterized by a rate of economic growth unprecedented in history. In the working of the system we see recurring critical periods. Nevertheless, growth in production is greater than the losses that may occur. The root of the problem is not growth as such, but the type of growth and the means by which it is achieved, and its effects. Let me say once again: the system is unjust. Some economists say this is unimportant. According to them, the system is not concerned with moral questions, but with whether it is efficient or not in the production of wealth.[5] If this is achieved, then despite its high social cost, the system is acceptable.

For whom? Certainly not for those who have to pay substantially and proportionately more to keep the system going, the poor, who bear the brunt of the system's injustice. They do so, first, in material terms. Indeed, the process of globalization manifests itself, among other things, in a levelling of prices at international level, though this is not accompanied by a like tendency to level wages. Those who are able to travel between the north and south of the planet can experience how the price range of the same product decreases in the market; the same merchandise is sold at very similar prices in different parts of the world. Of course there are exceptions, but these are explained by particular circumstances. However, this similarity in prices is not matched by similar wage levels. It is a fact that the value of work by the poor does not follow the tendency for prices to equalize in the market. This fact is more evident in the peripheral countries and in the points of the system linking the centre/inner circles with the outer circles or periphery. Because they have to pay ever increasing prices with wages that do not increase at the same rate, people become poorer.

Secondly, for just over fifteen years, we can observe a tendency for solidarity in public and governmental terms to decrease both at international and domestic level. Few countries agree to contribute at least 0.7% of their GNP to aid the socio-economic development of poor countries. It is true that, privately, there are social sectors which have tried to make up for this lack with voluntary contributions. Obviously, despite their generosity, this help is far from being sufficient. But without these contributions the situation of the poor would be much more difficult. However much energy the poor put into trying to survive, poverty and misery are increasing nearly all over the world. And to add to the gloom, this unconcern for the fate of the poor is also manifested in the domestic policies of most countries. The application of 'neo-liberal' policies and special economic adjustment plans leave the poor in a terrible position. Deprived of all care, it is not surprising that they are attracted towards various forms of violence and other illegal means of survival. Much of the responsibility for the development of these processes lies with the predominant tendencies of the system.

Thirdly, 'the irrationality of the system's rationality' translates into an ideology which fosters individualism and decreases the importance of the social and its values.[6] The only social 'thing' which appears to count is 'the market' (which, as we know, is very unsocial). This ideology promotes a dominant 'common sense', which leads the majorities, both in North and South, to accept very high costs and sacrifices, for the sake of the system, the market. It is a convenient ideology for those who administer the system's institutions and processes, but also an ideology supported by many of the poor, prepared to sacrifice themselves on the system's altars. At this point I believe it is possible to speak of a 'civil religion',[7] at whose sacred centre stands the 'totem' (because it is an human artefact) of the market.

Finally, this leads many to accept *exclusion*. They even accept that it should be affirmed through practices, policies and decisions which go against the poor. Unfortunately, the policies are often supported by the poor themselves, who are motivated by 'the desire to copy' and want to be like those who are successful and live in affluence and leisure: much more frequently than we would like to accept, the poor sacrifice others who are poor.[8] The tragedies which occurred in Ruanda and Sri Lanka, to cite two cases among many, are examples of what I mean. I repeat: it is a matter of copying. This means that the sacrifices imposed in accordance with the system's logic are the most unacceptable. They set an example inducing the least privileged to practise injustice and violence.

Life is threatened

Those who run this process and carry it on have maintained that it is the best possible system (and continue to do so). They also argue that the culture expressed by this system is the highest that human beings have constructed throughout history. As we have seen, it is a culture of domination, which frequently leads to great explosions of violence – while simultaneously proclaiming high ideals for humanity. The 'official' discourse of the system's culture claims it wants to defend life. Unfortunately 'the irrationality of its rationality' puts life in danger.

The logic of this system, which originated in Western Europe, has become the way the 'dominant class' operates today; it is not only Western but trans-national. Its outlines can be traced back nearly three thousand years. It began to be defined more clearly with the process of the 'Archimedization of nature',[9] which took place during the transition from the sixteenth to the seventeenth century. Francis Bacon (1596–1626), Galileo Galilei (1564–1642) and René Descartes (1596–1650) advanced rapidly along this path. Just as the Christian West conquered and colonized infidels and pagans, so man was called to conquer nature.

A fundamental dualism was established between the subject, capable of exercising knowledge about an object, and this object, thought of as external.[10] The subject, separated from other forms of life, felt called upon to dominate nature. Biblical-theological justifications were even given for this.[11] The subject, more specifically, 'man', placed himself in the centre of the system, as well as seeing himself as its crowning glory.

Nature became man's dominion. Despite evolutionary theories, which see the human species as one more link in the chain of life, methods of scientific investigation were developed on the basis of abstractions, most of which were measurable, because in this way it is easier to manipulate the phenomenon of life. But life in the abstract has no spirit. It can be reduced to various quantities: metres, kilometres per hour, amounts of power and energy. This is a measured life, life without mystery, with no sacramental dimension. Hegel wrote in his *Phenomenology of the Spirit* that 'nature has no history'. On the other hand, the Spirit does have one: it realizes itself through its historical journey. However, for systematic thought, nature has little to do with this. It is there to be used and profited from, as an instrument. Consequently, the subject was unaware of his natural limits. Human *hybris* has no objections to invading the environment in which we live. Economists, whose main concern was the generation of wealth and well-being, strove to elaborate theories which would result in the maximum optimization of economic growth, a growth which accelerated unbelievably during the course of the past fifty years, pressing on towards ever higher and more distant goals. Through this kind of growth, damage was done to the environment. It was irresponsible and excessive.

Life, according to contemporary biology,[12] cannot be treated this way. It requires investigators to place themselves within the context of the life process, to which they naturally belong. The works of the human spirit, works of life, cannot be isolated from nature. Therefore we can say that culture and love are part of life, not only in personal and social terms, but also biologically.[13] In our time, cultural development which is becoming aware of the system's irrationality, is urgently demanding an end to the dualistic focus inherent in this culture of domination, which leads to the manipulation of life and produces a disastrous ecological imbalance.

The system is wrong and it is imperative to correct it. In order to do so, we must confront important challenges. First, we must break the spell of the system. It is not the only system possible. Secondly, there is a challenge to our imagination: it is possible to imagine a different system, which would be more socially just, more economically fair, and ecologically viable. Thirdly, we also face a cultural challenge: how can we place science and technology at the service of the most needy, without at the same time damaging the environment? Fourthly, there is an ethical challenge: to behave responsibly towards future generations. If we must have growth, let it be sustainable.[14]

The gift of life which we receive from God has inestimable value. The systems that we set up must not go against the meaning of creation. Are we prepared to pay the cost of this? Values must not only be stated but sustained. For this to happen we have to translate values into actions, programmes and decisions which affect the economy and society. This means that we must work against the system within the system. This is a paradox, but one which we think corresponds to the mystery of life.

Translated by Dinah Livingstone

Notes

1. Franz Hinkelammert, *La irracionalidad de la racionalidad dominante*, San José de Costa Rica 1994.
2. The concept of 'longue durée' to explain historical processes was developed by Fernando Braudel. Cf. his *La Méditerranée et le monde méditerranéen à l'époque de Philippe II* (2 vols.), Paris 1966.
3. The name given by Immanuel Wallerstein, *The Modern World-System, II*, New York, London, etc. 1980.
4. Take, for example the 'Mexican crisis' and its impact on the 'newly emerging markets'.
5. This is the classic position of Milton Friedman, among others.
6. Remember Margaret Thatcher saying 'There is no such thing as society'. We find a similar viewpoint in the 'Reaganomics' that ruled the USA between 1981 and 1992.
7. 'Civil religion' according to the meaning given to the term in Rousseau, *Du Contrat Social*, Paris 1976, 427.
8. Cf. René Girard, *La Violence et le Sacré*, Paris 1972.
9. Cf. Antonio Banfi, *Galileo Galilei*, Milan 1961.
10. Hans Jonas, *The Phenomenon of Life*, New York 1966.
11. Based on crude interpretations, informed by the dominant cultural tendencies in the West, of passages such as Gen. 1.1–2.4 and Gen. 2.
12. See H. Maturana and Francisco Varela, *El Arbol del Conocimiento*, Santiago de Chile 1990.
13. H. Maturana and Gerda Verden Zöller, *Amor y Juego: Elementos olvidados de lo humano*, Santiago de Chile 1993.
14. Paul Abrecht (ed.), *Faith and Science in an Unjust World*, Geneva 1980.

Ecological Consciousness in Amazonia: The Indigenous Experience

Berta G. Ribeiro

I. The environmental crisis

The profound techno-scientific changes of the present age led the French philosopher Félix Guatari to investigate, in *Les trois écologies*, the extent to which they affect the inequalities among nations and the antagonisms between the developed and underdeveloped world. The information revolution and other technological advances threaten to make human work potentially dispensable. Guatari gives the example of the Fiat works, in which the workforce has been reduced from 140,000 to 60,000 operatives, while productivity has increased by 75 per cent. Will this dispensability of human labour lead to idleness, marginalization, neuroses and criminality, Guatari asks, or to a creative reinvention of our life-styles?

The First World, despite its undeniable prosperity, faces the danger of nuclear accidents such as Chernobyl, a consumerism requiring unlimited energy, acid rain, smog, contamination of its waters or the poisoning of the soil by toxic wastes, the mountains of waste produced by packaging, to name but a few of its problems. The situation in the Third World is more dramatic, with three-quarters of the population living beyond the reach of the advances made by the first industrial revolution: general free availability of first-grade education, access to paid work, to decent housing and to conditions of public health and hygiene such as to assure them of physical and mental health.

In the Third World, the crisis is, then, social and environmental. Socially, it is manifest in the shape of absolute poverty: the explosive growth of cities, loss of minimum standards of quality of life. From the environmental standpoint, the tropical regions of the Third World have equally serious problems. Its forests, which used to cover 20 per cent of the earth's surface – in Latin America, Africa and South-East Asia – now occupy scarcely 7 per cent, with the majority of these – 57 per cent – in South America. The scale of deforestation is alarming: every two seconds an area the size of a football pitch is cut down; every day an area the size of a city such as Bogotá (30,000 hectares); every year an area the size of a country like Guatemala (108,000 sq. kms.); every ten years that of Peru (1,285,000 sq. kms.). If destruction continues at this pace, there will be no tropical forests left on earth by the year 2080.

In this context, Amazonia emerges as one of the main victims of a process of devastation, started in 1970 on the pretext of occupying 'empty spaces' and relocating populations from other areas. Instead of progress, what this has brought is 'developmentalist chaos', financed to a great extent by outside capital, particularly in the sphere of mining. But mining is not all: hydroelectric dams flood huge flat areas, drowning species and dislocating populations, but there are no means of installing transmission lines. Clear-felling and burning, subsidized with tax incentives, destroy riches worth far more than the value the land can ever produce. Indigenous territories are invaded wherever they might contain something of commercial value, their inhabitants reduced to beggary or exterminated. The cultures of small-scale extractors, peasant farmers and fishermen are overrun, while their values are overturned by the price put on their land. The picture is made worse by the re-emergence of endemic diseases such as malaria – including an urban form, as in Manaus – Leishmanaisis, tuberculosis, and the re-surfacing of long-banished plagues such as cholera.

Forest clearance and the sowing of grazing for cattle leads to the erosion and toxification of the soil. Prospecting for gold pollutes and silts up the rivers; its profits are diverted by smuggling and find their way into the drugs trade. Now that the potential for employment offered by the opening of mines, the building of roads and dwellings and the construction of hydro-electric dams, which once mobilized a workforce running into millions, has been exhausted, there is no work for the workers; they are laid off and drift to the urban centres, creating huge slum areas. Manaus today holds 70 per cent of the population of Amazonia, an area three times the size of France; Bõa Vista has 80 per cent of the population of the state of Roraima.

Just like the rise and fall of the rubber industry in the last century, which enriched only a few, the present-day pattern of occupation of Amazonia is proving perverse in two respects. First, it destroys non-renewable natural resources; second, it excludes most of the population, both indigenous and immigrant, from the profits generated. The immigrants were encouraged to move to Amazonia by the government, which now either cannot or will not give them access to land or offer them a standard of life that measures up to what they had where they first came from.

In social terms, therefore, development cannot be judged merely from an economic viewpoint, but must also be seen as an ethical and ecological problem. Economic changes and technological advances are rendering the poor part of humanity removable, or economically superfluous. And yet, to the contrary of what might appear, the poor should not rebel against science and technology, since overcoming their poverty depends on scientific and technological knowledge, even if the transfer of this to them has to be made under commercial patterns.

Collective knowledge was handed on without any charge. We know that as a result, the aboriginal populations of the Americas bequeathed to humankind the majority of the plants on which we feed – one thinks of potatoes, tapioca, millet, tomatoes, sunflowers, almonds, cocoa, *verba mate*, industrial plants such as rubber, and so many others. The tribal populations of the Americas are still guardians of genetic information that, biologists say, could become much more valuable than raising cattle. Establishing property rights on natural information – understood as the value resulting from lack of human manipulation – would be one way of recompensing the heritage preserved for thousands of years by the Indians, and for hundreds by the other forest-dwellers. In other words, paying royalties to the owners of virgin habitats, which are of most interest to the pharmaceutical laboratories, is one way of ensuring their preservation.

In the following pages, I present some case studies. The first is from my own experience with an indigenous group, the Desâna, who speak Tukâno and come from north-east Amazonia, on the border with Colombia; next, Darrell Posey, with the Ka'apor, of the Jê linguistic family, from the Xingu region; then William Balée, with Tupi-speaking tribes from Maranhão and southern Pará; finally, other examples illustrating points of ecological anthropology. These subjects come under the general heading of ethnoscience, understood as defined by Cardona (1987, 11) as: 'The totality of local knowledge of particular habitats, the manner in which this is presented through the filter of such knowledges and not directly through observation by Western naturalists. Within this ethnoscience, we can speak of an ethnobiology, ethnobotany, ethnozoology and other possible subdivisions.'

II Soil classification: the Desâna

A clan of seventy-five Desâna Indians inhabits the village of São João, on the middle reaches of the Tiquié river, which flows into the Uaupés, itself a tributary of the upper Rio Negro in Amazonia. The areas irrigated by blackwater rivers, classed as oligotrophic (relatively poor in plant nutrients), are considered the poorest in nutrients of Amazonia. The soil is whitesand, highly acidic (4pH), and the rainfall extremely high (over 140 inches annually). Compensation for these conditions is provided by a 'recycling system' made up of a 'carpet of roots above ground, some 20 cms deep, which traps the fallen leaves, facilitating their decomposition through interaction with fungi and mycorrhiza [symbiotic association of fungus and roots] which effectively prevents loss of nutrients' (Moran 1991, 172). The tribal groups, such as the Tukâno-speakers, who live in hierarchically placed villages along the blackwater rivers, cultivate few species of vegetables, concentrating on wild manioc (cassava, tapioca – *Manihot esculenta*), which adapts best to this ecosystem.

The Desâna classify the soil according to its composition, whether or not it is flooded at some season of the year, and the type of vegetation it supports. To do so, they have developed their own taxonomy, knowing exactly where to find the plants they need for their artefacts, remedies and other uses, and the terrains that are best suited to cultivation.

They distinguish four eco-zones in their region: *terra firme, caatinga, igapó* and *manquezal* (moving from driest to swamp). Only *terra firme* can be used for cultivation. It is in turn sub-divided into four types: 1. sand, sub-divided into ordinary and black; 2. clay, also ordinary and black; 3. yellow loam, best for cultivation, but rarer; 4. clay, damp and light, also better for working, but occurring only in small patches along the river Tiquié. They learn the type of soil by digging it.

Land not used for cultivation is classified according to the height of the trees it supports and its surroundings: aquatic, periodically flooded or dry. *Caatinga*, whose trees are not as tall as those of *terra firme*, is classified as dry. These areas are reserves of fruit trees and primary products for artefacts. The most important are: the great Bacaba palm (*Oenocarpus bacaba* – a source of palm oil), the Sorvo and *sorvo do pará* (*Couma macrocarpa* and *Couma utilis* – sources of coumarin or couma rubber), and various types of cypress used to make fish traps. The *caatingas* also contain a great variety of medicinal plants differing from those found on *terra firme*. The Desâna distinguish smaller patches within the *caatinga*, which they call 'small *caatinga* terrains'. These are recognized by the presence of large clumps of trees bearing edible fruits and reeds used for fishing.

Another phytogeographical division noted by the Desâna is the *igapó*. This formation occurs on river banks, and is inundated during floods. The soil is clay with a small admixture of sand and cannot be used for cultivation. This division, however, contains the greater part of the flora on which fish feed, as well as primary products for artefacts. Tolaman Kenhíri (or Liuz Lana), my chief Desâna collaborator, collected ten plant species from the *igapó*, those he considered most important for fish that feed on fruit (see Ribeiro and Kenhíri 1992 ms).

According to Chernela (1986, 242), of the forty-one fruit-bearing species that grow on the river banks, on whose fruits fish feed, twenty-seven grow only in the *igapós*, while fourteen grow on *terra firme*, of which ten also grow in the *igapó*. These contain great concentrations of Acai palms (*Euterpe* sp. – a source of palm hearts), native to Amazonas, of Inga (*Inga* sp.) and other trees of which fish eat the fruit. The most prominent are the Jauary or Javary palms (*Astrocaryum jauary*).

In the *manguezal*, a permanently wet region, swampy, with blacksand soil, no use for cultivation, the Desâna also find plants useful for their artefacts or with edible fruits. These include two species of arum (*Ischnosiphon* sp.), the main material for baskets, several species of yellow cypress, Bataua, Acai and Ita palm (*Jessenia bataua* and *Mauritia flexuosa* – sources of sago, oil and wine), the latter in large clumps, as well as the tree that provides resin, the Brazilian Elemi tree (*Protium* sp.). Most of these species also occur on *terra firme* (sand or clay), where they are more easily found, as these areas are more frequented by the Desâna.

The collection and preparation of plant extracts used for building houses, making canoes and paddles, baskets, ceramic and other artefacts, amounting to a total of seventy-four plants, most of them classified scientifically by the National Amazon Research Institute (INPA), show that most (forty-one) come from *terra firme*, with twenty-one from secondary growth, sixteen from the *igapó*, four from the *caatinga*, and four from the *manguezal*. In many cases, the same plants grow in more than one of these ecosystems (see Ribeiro 1990 ms).

III Desâna horticulture

The Desâna, like the other indigenous groups of the upper Rio Negro, carry out cultivation based on slash and burn. The cutting down of virgin forest and secondary growth is determined by the annual cycle of the constellations. These signal the approaches of heavy rains, which have the same names as those given to the nineteen constellations recognized by the Desâna. In the intervals between these, when various fruits are harvested, including *canistel* or *sapote*, peach nuts and *uvilla* (*Pouteria caimito, Bactris gasipaes, Proqueima sericea, Porouma cecropiifolia*), there are summer and winter periods when burning is carried out. At least seven days of strong sun are needed to be able to burn the wood cut down in the clearances. The ashes and charcoal fertilize the soil (see Ribeiro and Kenhíri 1987; Ribeiro 1990).

Each nuclear family prepares a plot of approximately one hundred square yards each year. Until harvest time, they subsist on the plot planted the previous year and two others, planted two and three years back. After this period, the invasion of weeds that absorb the nutrients in the soil render the plots unproductive. Besides wild manioc (cassava, tapioca), which occupies some 80 or 85 per cent of the plot, they plant *cará*, sweet potato, a small potato used to make the fermented liquor generally called *caxiri*, several varieties of fruit tree and, on some types of soil, water-melons, peppers and medicinal herbs.

Except for the cutting, agricultural work, and the processing of the manioc, is done by women. Nevertheless, besides helping the women with digging and fertilizing, and sometimes harvesting, the men cultivate certain species practically for their own exclusive use: these relate to occupations such as smoking, drugging fish in still waters, and tying basket-work and arrow heads.

A new plot can be replanted once, including with manioc, which is a plant little subject to pests, grows in poor or acid soils such as those in north-west Amazonia, and can be harvested after six months. Other advantages of this root crop – compared to grains, such as millet – are that the tubers can be stored for long periods in the ground, it requires no burning or general clearance of the land, and its unit/labour and unit/area productivity is much higher, as is its calorie count. Despite its low protein content, the indigenous populations, including the Desâna, whose diet is made up of manioc to around 70 or 80 per cent, are physically strong and show no signs of nutritional deficiency. Another advantage of wild manioc is the great number of food products that can be extracted from it. Among the Desâna of São João alone, I counted twenty-seven dishes made with this tuber as a basis – eight grades of pulp, five types of flour, seven unfermented drinks and eight fermented.

According to Chernela (1986, 157), the Wanâno, a Tukâno group from the Uaupés river, grow 137 cassava cultivars. Dufour, cited by Moran (1191a, 368), counted one hundred cultivars of this tuber in another Tukâno group from Colombia, the Tatuyo. Exogamic marriage makes the women spread through various villages and, as they visit relatives, they exchange cultivars and experiences with them (ibid.). Kerr (1986, 168) obtained forty Desâna names for this tuber, with Kenhíri, who pointed out the characteristics of leaf, root and core that distinguish each cultivar. These figures represent the largest tally of manioc cultivars known by an indigenous group: Carneiro (1983, 99) obtained forty-six among the Kuikúro and Boster (1983, 81), one hundred among the Aguaruna and Huambisa, Jivaro sub-groups of Peru.

The cultivation of manioc, unlike millet and beans, which do not flourish in poor soils such as those of north-west Amazonia, has made it possible for tribal groups to live in the area for centuries. Agricultural exploitation is possible here only through sensible management of the forest. In effect, as Moran (1991a, 368) notes: 'In blackwater ecosystems, regeneration of the original vegetation can take over a hundred years [Uhl, Clark & Clark 1982; Uhl 1983]; except where cultivation is for a short time and unless special steps are taken to accelerate recolonization by native tree species.'

IV Protein sources

Like the soil of the upper Rio Negro micro-region, its hydrographic system also suffers from extreme acidity. The black waters are chemically poor in soluble salts and their copper-brown coloration prevents the penetration of the spectrum of sun's rays necessary for photosynthesis. There is no plankton or floating plant-life on which fish can feed. Many species obtain nourishment directly from leaves, flowers, fruits, insect larvae and adult insects, fungi, bacteria and other micro-organisms. So the Indians never cut down the vegetation overhanging the river banks, showing that they know the interdependence between the land and water systems on which the maintenance of fish stocks depends.

The poverty of the icthyological fauna and these feeding habits of the fish have determined the development of an elaborate technology for catching fish through the use of basket-work fish-traps and other refined techniques. Fishing with any of these implements, to be fruitful, has to be based on the knowledge the Indians have of the biology of the fishes.

They classify them on a variety of criteria. One of these is whether their feeding habits are nocturnal or diurnal – a criterion also applied to animals and birds. A second classification used by the Desâna is by diet: there are icthyophagic fish, which eat the flesh of other fish, and carnivorous fish, which feed on crayfish, spiders, grasshoppers and any type of insect that falls into the water. There are also fruitivorous fish, which eat the fruit, seeds and stones of trees that grow on *terra firme* and, mainly, in *igapó*.

An important classificatory differentiation of fish is on the basis of their reproductive systems. The criterion used is to concentrate on whether their eggs are grouped together or broadcast. Fish that spawn while migrating are classified as belonging to the *tanini* group. Those that spawn without migrating, the fish that remain in the stretch of the river Tiquié used by the Desâna clan that we studied, are grouped under a generic heading which might be translated as 'enclosed egg'. A third group called 'left egg' designates fish that spawn in hidden places, under the big roots of trees.

A fourth category of fish is classified by the habitat in which it is based: lake, river or swamp. The lake-based fish are endogenous – they do not migrate. There is also a distinction made between fish that leap and those that only swim, and a further one by the depth of water they inhabit, differentiating those that swim on the bottom, at half depth, or on the surface. The passing of a shoal is identified by the period of the year in which it migrates, according to the calendar of the constellations, and also by the greater turbulence of the water and by listening to the tumult, or 'music', of the fish as they spawn. The Desâna believe that the spawning grounds have not changed since the dawn of time; they can be identified by trees and bushes forming a dense overhang on the river bank, or where projecting tree roots produce a similar effect.

Nets were introduced on the river Tiquié in 1979 and now form the main fishing implement. Despite this, the Desâna consider that none of their traditional traps has been rendered obsolete. They are still used by those too poor to afford nets. Despite gloomy prognostications, nets have not exhausted the fish stocks, as the fry pass through their meshes.

Each stretch of river has its endogenous fish, so much so that the Indians say they do not know how to fish in regions where they prospect or other distant parts. The fish local to São João, on the other hand, are minutely known and named. This is because of the abundance of species and low numbers of each species in the blackwaters: Golding and others (1988, 109) reckon there are over 700 species of fish in the lower Rio Negro, most of which are diminutive, under two inches long. Even these tiny fish are recognized and named by the Desâna, who can also distinguish between those that will grow and those that remain small.

According to the Desâna, the constellations that signal the timing of rains and droughts, influencing times of slashing and burning, and the running of shoals of fish, spawning or not, also regulate the harvesting of certain edible invertebrates: ants, termites and caterpillars. Tied to the climatic changes associated with constellations, the harvesting of edible invertebrates is still an important source of animal protein. These include *cupim* (*Conitermes* sp.), *saúva* (*Atta* sp.) and larvae of butterflies which eat the leaves of the Micranda rubber tree (*Micranda spruceana*), *cunuri, tururi, japurá, inga* and *acariúba* trees. The 'maturation' of these larvae occurs between July and September, with the greatest concentration in August.

The detailed Desâna classification and taxonomy of these various species shows the deep knowledge the Indians have of these edible invertebrates and their dietary value. A study by the US anthropologist Dufour (1987), carried out among the Tatúyo Indians, Tukâno-speakers from Colombia, proves that protein from these invertebrates makes up 12 per cent of the annual animal protein intake for men, and 26 per cent for women. Besides those mentioned, they eat beetle larvae and wasp pupae, but these are not linked to the constellation cycle.

Hunting is of less importance for the Desâna and other Tukâno-speaking groups than fishing. Game has traditionally been provided by the forest-cultivating Makú, better adapted to dry-land vegetation, who exchange game and forest fruits for manioc flour and indigenous and exogenous artefacts with the river Indians. Silverwood-Cope (1972, 96) relates that, during his period of observation, 40 per cent of the meat hunted by the Makú was traded with river Indians.

V Forest and enclosure management

Cultivation of the tropical forest can be classified as a 'vegetable civilization', like those of other regions of America which attach more importance to plant than to animal domestication. The historiography of the continent has listed more than one hundred vegetable species originating in America and taken and acclimatized by Europeans in all quarters of the world. They fall under the headings of alimentary, medicinal, handicraft, fuel, aromatic, dyeing, flavouring, ornamental, emblematic, gum and resinous, used for all sorts of purposes (Martinez 1990). The following are examples of studies made among Amazonian indigenous groups on enclosure and forest management.

Posey (1986) demonstrates that both forest and savannah are known and actively manipulated by the Ka'apo. They create artificial bush cover, which they call *apetê*, meaning 'made', increasing the biological diversity of areas of creeping vegetation. They plant by making shallow depressions in the earth, which hold rainwater, filling them with straw mixed with pieces of termite-mounds and discarded ants' nests; they also put in living termites and ants; these fight, and leave the plant shoots alone. The dead insects decompose and add nutrients to the soil formed in this way.

Mounds of soil are thereby formed, up to six feet across and two feet deep. In time, they grow, and become 'resource islands' in the middle of the savannah. These islands produce fruit trees, which attract wild-life, shade-giving trees, firewood and creepers that give drinking water: all these are 'semi-domesticated', transplanted from enclosures to form new ecological niches. The same practice can be seen along the trails the Ka'apo make through the forest on their seasonal hunting trips (Posey 1986, 177).

Balée (1989, 4, 11–12) and others consider the 'Indian blacklands', as an example of which they give the site of the city of Santarém, in Pará, 'anthropogenic', meaning that they have been built up by generations of Amerindians through the accumulation of organic matter from their village terrains. Balée also raises the hypothesis that concentrations of certain palm trees and other useful species resulted from manipulation of ecosystems in order to create niches favourable to human existence. This is a process of intentional modification of habitat to stimulate the growth of animal and human communities. By reason of this, Balée considers that the current notion that the aboriginal populations of Amazonia are adapted to the virgin forest should be turned round: this adaptation is the result of 'agro-forestation' owing to management of the environment by extinct peoples. He states that in the '*cipó* forests', inhabited by two Tupi groups which he studied, species foreign to this biotype were found. Such are Brazil Nut trees (*Bertholletia excelsa*), Bacuba palms (*Orbignya phalatera* – a source of palm kernel oil), 'wild' cocoa (*Theobroma* sp.) and Inga (Balée 1989, 12).

Capoeiras, old clearings left fallow after two or three years of cultivation, are not abandoned land. They continue to produce crops for several years and furnish shoots and thorny plants on which animals feed. They are therefore 1. banks of germoplasm, that is, hybridization and seed reserves; 2. residual orchards of various fruit trees; 3. 'game reserves' in which planting to attract game has made it easier to find rabbits, peccaries and other animals than in the virgin forest (Posey 1986, 175).

The growth of invasive plants is tolerated in order to produce fresh burnings, the ashes from which fertilize the soil, while the smoke destroys pests attached to new shoots, and the first rains produce grasses which attract herbivores. A headcount of animals hunted in three months, during the dry season, by the Ka'apor of Maranhao, shows that 29.3 per cent of the total came from clearings made in this way, an extraordinary percentage in view of their relatively limited extension (Balée 1984 ms, 225).

The advantages of agricultural techniques that are extensive (as opposed to intensive, mechanized), polyculture (as opposed to monoculture, of cash crops for export), can be summarized as follows, according to Meggers (1988, 42–4).

– they maintain the fertility of the soil by not totally eradicating its plant cover;
– the clearing of a small plot of ground – around a hectare per family – and its temporary utilization minimize the time during which the ground is exposed to the heat of the sun and heavy rainfalls;
– planting of different species, of varying heights, reduces the effect of storms and pest propagation;
– geographic dispersal of cultivated plots ensures the preservation of vegetable and animal species in 'natural corridors' separating the clearings and acting as ecological refuges;
– in small-scale burning, the ashes and rotting of branches and trunks left unburnt return to the soil the nutrients needed for feeding new shoots.

Because of the contribution which itinerant agriculture makes to the proliferation of game animals, Linares (1976) considers the clearing/hunting interaction a result of the domestication of animals in tropical America.

Compared to this agricultural management, what is happening in the 'civilized' occupation of Amazonia is the clearance of 10,000 hectares in a single year and at a single stretch to provide grazing lands. The grazing lasts only two years and produces the derisory amount of thirty kilos of beef per hectare. This means that one head of cattle eats the equivalent organic matter to that consumed by an indigenous family, with the added damaging effect of destroying the soil and, as a result, altering the climate and the water system through irrational clearing of virgin forest.

VI Conclusions and perspectives

Indigenous management of Amazonian eco-systems, as outlined in the few examples above, shows that they have a clear understanding of the habitat in which they live. They recognize ecozones associated with particular plants and animals, whose biology and behaviour they know in detail. Plants and animals, including invertebrates, are used by them for food, medicine and artefacts.

There has been a great deal of misunderstanding of itinerant agriculture. In fact, it is the only system adapted to the characteristics of the eco-systems of the wet tropics. The deliberate dispersal of areas of cultivation and the small extension not only prevent exhaustion of the soil and fauna and encourage regeneration of both, but also minimize the problem of pests, conserve the productivity of the soil and thereby ensure maximum advantage from natural resources.

The new outbreak of 'progress' not only threatens the survival of indigenous groups, the other 'forest peoples' and the Amazon rain forest itself. Government interventions, far from lessening the sufferings of the native populations, will only tend to exacerbate them. This is what Hill and Moran maintain, referring to north-west Amazonia: 'The eco-system of the Rio Negro can only support populations far smaller than those that national bodies wish to settle there in connection with the geopolitics. Concentration of persons in large communities cannot be supported by local production processes, given the oligotrophic conditions' (1983, 134).

This is the great threat facing the Indians and non-Indians of Amazonia. In the case of the former, it has been possible to meet it by socio-economic organization, limiting the population growth and dispersing it over a vast area. Furthermore, government action has resulted in some technological advances, benefitting the native populations without requiring them to become directly involved in the market economy, where they could occupy only the lowest rungs. These conditions, however, seem to be evaporating, making it difficult to predict what might happen in the near future.

What is certain is that, given the conditions of interaction, it is impossible to envisage complete sovereignty for the indigenous communities. It is inadmissible, however, for their territories to become the object of 'ecological colonialism', in the expression coined by Stefano Varese referring to Peruvian Amazonia. He says: 'Among the various forms of colonialism which Peruvian society has carried out and is carrying out in the Amazonian jungle, there is no doubt that territorial occupation and exploitation holds a prominent place. This is above all because of a national outlook that sees this part of the territory as land for conquest: land for extraction and not for reproduction; a sort of internal colony or "third world" which can be sacked for the sake of the ruling economic and social system.' He concludes that, like the forest, the native inhabitants are regarded as enemies to be eradicated. This is because governments that have not faced up to the urgent need for agrarian reform have pushed the dispossessed rural masses to the outlying areas.

This view is shared by Fernside (1990, 34) in his enquiry into what development plans are for. He answers: 'However big Amazonia may be geographically, it is not capable of resolving the problems of other regions, such as the lack of effective agrarian reform, which is largely the cause of the present migration to the region. Such problems can be solved only in the regions in which they originate.'

A complementary argument is provided by Balée (1993, 393). Citing Norman Meyers, he foresees that 'more than 17,000, or nearly half the 34,200 plant species endemic in these "hot places", will be extinct by the turn of the century. This number represents more than 13 per cent of the vegetable species of all the world's tropical forests.' He adds that 350,000 animal species will have become extinct in the same period. These 'biotic depletions' will be the result of the actions of state bodies and not of the Amazonian Indians or of the human race in general.

What hope remains is linked to increased consciousness of the ecological problems of Amazonia, umbilically linked to protection of the indigenous populations and other strata of the population who live in harmony with nature. The movement in defence of Amazonia, which has spread world-wide, is echoed in wide sectors of public opinion. The political and economic powers with an interest in the subsidized widening of areas for agro-pastoral, logging, mineral and hydro-electric exploitation cannot ignore this clamour. Like it or not, they will be forced to face up to demographic pressure on *latifundios* and the cities of the south and south-east (of Brazil) and to the historical challenge to create ways of managing the Amazonian forest that will be technically advanced, economically viable and non-polluting and socially just.

If this tendency is to coalesce and develop, a space has to be made for the Indians and other categories of 'country people' within the nation's society. In this space they must be neither subordinated nor patronized, but seen as active and specialized producers, capable of redeeming the historical patrimony of national identity. A space has equally to be made for rethinking economic development in such a way as to provide a better quality of life and a sharing in the nation's wealth for the greatest number.

Translated by Paul Burns (with thanks to Dr Mary Clare Sheahan at the Royal Botanic Gardens, Kew, for help with botanical identification)

Bibliography of works cited

Balée, William

1984. *The Persistence of Ka'apor Culture*. PhD dissertation, Columbia University, New York, unpublished.

1989. 'The Culture of Amazonian Forests' in D. A. Posey and W. Balée, eds., pp. 1–21.

1993. 'Biodiversidade e os indios amazônicos', in E. Viveiros de Castreo and M. Carneiro da Cunha, eds., *Amazônia: Etnologia e História Indígena*, São Paulo, pp. 385–93.

Boster, James

1983. 'A Comparison of the Diversity of Jivaroan Gardens with that of the Tropical Forest', *Human Ecology* 11.1, pp. 47–67.

Cardona, Giorgio R.
1987. 'Visión del Mundo Natural', *Hombre y ambiente. El punto de vista indígena*, Vol. 3, Quito, pp. 9–45.

Carneiro, Robert
1983. 'The Cultivation of Manioc among the Kuikuró of the Upper Xingu', in R. B. Hames and W. T. Vickers, eds., pp. 65–111.

Chernela, Janet M.
1986. 'Os cultivares de mandioca na área do Uaupés (Tukáno)', in D. Ribeiro (ed.) and B. G. Ribeiro (co-ord.), Vol. 1, pp. 151–8.

Dufour, Darna L.
1987. 'Insects as Food: A Case Study from the Northwest Amazon', *American Anthropologist* 89, pp. 383–97.

Fearnside, Philip M.
1990. 'Usos predominantes de la tierra en la Amazonia Brasileña'; in A. Anderson (org.), *Alternativas a la Deforestación*, Quito, pp. 363–93.

Goulding, M., M. Leal-Carvalho and E. G. Ferreira
1988. *Rio Negro. Rich Life in Poor Water. Amazonian Diversity and Foodchain Ecology as seen through Fish Communities*, The Hague.

Guatari, Félix
1990. *As Três Ecologias*. Campimas.

Hames, R. B., and W. T. Vickers, eds.
1983. *Adaptive Responses of Native Amazonians*. New York Academic Press.

Hill, Jonathan and Emílio Moran
1983. 'Adaptive Strategies of Wakuénai People to the Oligotrophic Rain Forest of the Rio Negro Basin', in Hames and Vickers, pp. 113–35.

Kerr, Warwick E.
1986. 'Agricultura e seleções genéticas de plantas', in Ribeiro and Ribeiro, Vol. 1, pp. 159–72.

Linares, Olga
1976. 'Garden Hunting in the American Tropics', *Human Ecology* 4.4, pp. 331–49.

Martínez, Miguel Angel
1990. *Contribuciones Latinoamericanas al mundo. La utilización de las plantas en diversas sociedades*, Mexico City.

Moran, Emílio F.
1991. 'O estudo da adaptação humana en ecosistemas amazônicas', in Neves, pp. 161–78.

1991a. 'Human Adaptive Strategies in Amazonian Blackwater Ecosystems', *American Anthropologist* 93, pp. 361–82.

Neves, Walter A. (org.)
1991. *Origens, adaptações e diversidade biológica do homen nativo na Amazônia*, Belém.

Posey, Darrell A.
1986. 'Manejo da floresta secundária, capoeiras, campos e cerrados (Kayapó)', in Ribeiro and Ribeiro, Vol. 1, pp. 173–88.

Posey, D. A. and W. Balée (eds.)
1989. 'Resource Management in Amazonia: Indigenous and Folk Strategies', *Advances in Economic Botany* 7, New York.

Ribeiro, Berta G.
1990. 'Etnobotânica Desâna. Plantas artesanais.' Paper to Second International Congress of Ethnobiology, Kunming, China.

Ribeiro, B. G. and Tolaman Kenhíri
1987. 'Chuvas e constelações. Calendário econômico Desâna', *Ciencia Hoje* 6. 36, pp. 26–37.
1992. 'Ectonictiologia Desâna', Paper to Estratégias Latino-Americanos para a Amazónia. Memorial da America Latina, São Paulo.

Ribeiro, Darcy (ed.) and Berta G. Ribeiro (co-ord.)
1986. *Suma Etnológica Brasileira*, Vol. 1, *Etnobiologia*, Petrópolis, RJ.

Sliverwood-Cope, Peter
1972. *A Contribution to the Ethnography of the Colombian Maku*, PhD thesis, Cambridge University.

Varese, Stefano
1970. 'Notas sobre el colonialismo ecológico', in A. Chirif (org.), *Etnicidad y Ecología*, Lima, 177–85.

Sacred Earth: Mesoamerican Perspectives

Sylvia Marcos

The sacred earth of the Nahuas

In Mesoamerican[1] cultures the earth is respected as soil. Today the idea we have of the earth, our home, is formed by a satellite photograph – the astronomer's blue planet, lost in the dark immensity of space. Mesoamerican peoples, however, drew their world-view from their embeddedness in the soil of a particular place. They regarded the layer of soil that supports all life on earth as living. This view is in sharp contrast to the contemporary view of the earth, in which that top life-supporting layer is subject to exploitation and destruction.

One day a gardener from a Nahuatl-speaking region was working in my garden. I told him to throw the dirt from a planter into the garbage bin. He answered with a shocked but polite, 'No, no, Señora, soil shouldn't be treated like garbage'. Another time, an Indian woman helping me at home heard me complaining about the dirt and dust blowing into the house. She chastized me, saying, 'Señora, you shouldn't speak like that of dust, because it is soil and soil is our mother, *la madre tierra*, who gives us food.'

The implicit ecological dimensions associated with Mesoamerican views of the earth and soil were brought home to me by these comments. What my assistants were reacting to was my acquired modern concept of soil as inert matter that can be discarded like refuse or complained about like an intrusive pollutant. They gave me a vivid lesson in ecology from the traditions of my own background.

In nearly all agrarian cultures the earth is sacred. Exactly how this sacredness is expressed and what forms it takes vary from one particular location to another. The unique features of the sacrificial sacredness of the Nahuas or Aztecs arise from their cosmovision. Many elements of the

Mesoamerican concept of the cosmos were often expressed in metaphor, a dominant mode of expression in Aztec culture.

Aztec cosmovision

Mesoamerican concepts and understandings of the earth differ radically from standard modern perceptions. Sources speak of the earth as a disc floating on water, a rabbit, an iguana or alligator (Cipactli) with a ridged back.[2]

Leon-Portilla writes:

> The surface of the earth is a great disc situated in the centre of the universe and extending horizontally and vertically. Encircling the earth like a ring is an immense body of water (*teo-atl*, divine water), which makes the world *cem-a-nahuac*, 'that-which-is-entirely-surrounded-by-water' (Leon-Portilla 1990).

In anthropomorphic representations, the earth was a body with eyes, mouth, hair and a navel. Earth itself is variously a womb, mouth and bowels. Body imagery was transferred to the multiple levels of the cosmos. The centre of the earth was its navel, the trees and flowers its hair. Grass was its skin. Wells, springs and caves were its eyes. Rivers were its mouth and its nose was the source of mountains and valleys (Gonzalez Torres).

Earth was also a devouring monster. According to Nahuatl cosmology, Quetzalcoatl and Tezcatlipoca brought Tlaltecuhtli (lord of the earth, also called Cipactli – alligator) down from the heavens. Tlaltecuhtli was a mythical male-female monster that snapped and bit like a savage beast. The two gods divided Tlaltecuhtli, thus separating sky from earth (Gonzalez Torres).

In ancient Mexico, horizontal space was emphasized much more than vertical space, contrary to the vision of the sixteenth-century missionaries with heaven above and hell below. Tlalocan was the 'paradise' of Tlaloc, the god of rain, and was located to the east, rather than in the heavens. Hell is a cave in the forest in the contemporary Nahuatl version. These examples indicate the strength of the horizontal metaphor in relation to the vertical Christian one (Burkhart).

The Mesoamerican universe is divided into four great quadrants of space whose common point is the centre or navel of the earth. From this point the four quadrants extend out to the horizon, the meeting place of the heavens and the surrounding celestial water (*ilhuica-atl*). Multi-layered symbolism is implicit in the concept of the four directions of the world.

Contemplating the passage of the sun, the Nahuas described the cosmic quadrants from a position facing the West:

> There where it sets, there is its home, in the land of the red color. To the left of the sun's path is the South, the direction of the blue color; opposite the region of the sun's house is the direction of light, fertility and life, symbolized by the color white; and finally, to the right of the sun's route, the black quadrant of the universe, the direction of the land of the dead, is to be seen (Leon-Portilla 1990).

This is the Nahua image of horizontal space. Vertically, above and below the horizontal world (*cem-a-nahuac*), are thirteen heavens and nine underworlds. Above the upper worlds is the metaphysical beyond, the region of the gods. Ultimately above all is Omeyocan (the place of duality), the dwelling-place of the dual supreme deity, the originator of the universe.

Rivers, lakes, water holes, mountain tops, caves, forests and deserts all had gods and goddesses who ruled over them and required rituals. Mountains in particular had a complex role in mythology. Rituals for mountain gods were celebrated throughout the annual cycle, as were those associated with the *cenotes* (underground lakes). Contemporary Mesoamerican Indians have preserved some of the complexities of these rituals. Among the Cuicatec, for example, cave rituals, mountain ceremonies and waterhole rituals are quite important. This has also been observed among Nahua communities of the state of Morelos near Mexico City.[3]

All deities which symbolized aspects of reproduction, birth and death had earthly aspects. Symbols of vegetable or animal deities were earthly symbols because the earth was conceptualized as the primordial image of the generating and regenerating principles of life. Death was an integral part of these forces. The earth had both human and animal characteristics; it was both male and female, living and dead.

Centuries of observing the world and its workings from the macrocosm to the microcosm of the body itself produced Nahuatl thought with its distinctive characteristics of duality, fluidity and balance.

Duality

Bi-polar duality is ubiquitous in Mesoamerican concepts of the cosmos. Dualities such as life and death, good and evil, female and male, earth and sky structure the Mesoamerican world. Though opposites, these pairs were complementary. The feminine-masculine duality is a typical example. Feminine and masculine fused in one bi-polar principle. This dual unity was fundamental to the creation of the cosmos, its (re)generation, and sustenance. Cipactli, the mythic monster who was lord of the earth, was male and female. Both singular and dual, this principle manifested itself as representations of gods in pairs.[4] Many Mesoamerican deities were pairs of gods and goddesses, beginning with Ometeotl, the supreme creator whose name means 'two-god' or dual divinity. Dwelling in Omeyocan beyond the thirteen heavens, Ometeotl was thought of as a feminine-masculine pair. Born of this supreme pair, other dual deities, in their turn, represented natural phenomena (Andrews and Hassig).

Fluidity

The duality implicit in Mesoamerican cosmology was constantly in flux and never fixed or static. An essential component of Nahuatl thought, this movement gave its impulse to everything: divinities, people, objects, time, and space with its five dimensions (Marcos 1993). Everything in the world was perceived on a continuum; everything flowed between opposite poles. Not only did the deities participate in a duality that shifts between opposites such as good and evil, but all entities and forces played a dual role: ' . . . (F)rom the four pillars of the cosmos at the four corners came the heavenly waters and the beneficial and destructive winds' (Lopez Austin 1984).

Balance

Movement characterized the Mesoamerican universe, but that fluidity was always in balance. The continuous seeking for balance gives duality a constant plasticity, making it flow and impeding stratification. Fundamental to the maintenance of the cosmos, this 'fluid equilibrium' is always re-establishing itself, keeping all particular points of balance equally in constant motion. Accordingly, categories of opposites such as female and male, day and night, up and down, near and far, good and evil were always in fluid equilibrium. In every situation, the 'critical point of balance' had to be found in continual movement; it redefined itself from moment to moment, and was subject to the change and flux of the entire cosmos, oscillating and ever reconstituting itself.

This never-ending search for cosmic equilibrium presupposed collective responsibility. Maintenance of that equilibrium was the 'moral' duty of everyone in the community. Because Mesoamerican thinking was based on the concept of opposing dualities and the search for balance between them, achieving equilibrium required each individual in every circumstance constantly to seek the hub of the cosmos and co-ordinate himself or herself in relation to it. To maintain this balance is to combine and recombine opposites. This is accomplished, not by negating the opposite but rather by advancing towards it and embracing it, in an attempt to find the ever-shifting centre of balance.[5]

This concept of a bi-polar universe, fluid and shifting, yet balanced, permeated the perception of beneficial and harmful actions as well as of good and evil forces, giving them a non-static, non-rigid quality. It also affected ideas about the deities and their power to interact with the earth and its inhabitants.

Metaphors

In the Nahuatl universe, everything was endowed with material, spiritual, temporal and spatial qualities. Consequently, it was a metaphorically complex and allusorily sophisticated construction (Andrews and Hassig). Leon-Portilla has called the Nahuatl culture a 'philosophy and culture of metaphors'. Metaphorical language is found largely in prayers, rhetorical orations (such as the *huehuetlatollis*), songs (*cantares*), and incantations (*conjuros*). As the main means of transmission of an eminently oral tradition, they were often memorized. The visual metaphors in the codices of all Mesoamerican cultures are pictorial representations of their cosmos.

In the Nahuatl world, metaphors were cultivated as the highest and most valued means of expression. Through metaphors, the Nahuas expressed their vision of the earth and the divine forces that affected it.

Immaterial, non-phenomenal things were not set off from the material world but were continuous with it, integrated into a single conception of reality. However, a tentative distinction can be made between 'physical' and 'ethical' metaphors: while there were special metaphors for the physical conception of the earth and for its position within the cosmos, other metaphors reveal the relationship that the Nahuas maintained with the earth and life on it, i.e., the moral perspective that guided them.

Moral dimensions of the Nahua earth and metaphors for it

The Nahua perspective on the earth is a moral one. The earth is a 'slippery' place (Burkhart), and the moral prescription is that one must act very carefully in all circumstances. One must live according to the guidelines established by the ancestors (Marcos 1991).

The earth is not a place of happiness. However, though it is primarily a place of effort and strain:

> . . . so that we would not die of sadness, our lord gave us laughter, sleep and sustenance, our becoming strong, our growing up; and moreover, earthliness (sexuality), in order that people go on being planted (Sahagun, translation by Burkhart).

The earth was above all a perilous place. The word *tlalticpac* synthesized many of the physical and moral meanings of earth and soil. It is formed by the substantive *tlalli* (earth) and the suffix *icpac* (on, above). However, its meaning is not just 'on earth', but rather 'on the point or summit of the earth', referring to a point of equilibrium on its crest and suggesting a narrow path between abysses. One linked oneself with the earth by acts of *tlalticpacayotl*, 'earthliness', which included, but was not restricted to, sexual activity (Burkhart).

Ancient chronicles are full of references to the relation between soil and 'sexuality' as well as between soil and moral matters. For instance, the grandmothers in the discourses recorded by Sahagun say that 'our bodies are like a deep abyss' (Marcos 1991). The *huehuetlatolli*, moral precepts of the parents spoken to their children, refer to the danger of earthly existence in these words:

> My daughter . . . Here in this world we walk along a very high, narrow and dangerous path like a very high hill with a narrow path along the top of it, and on both sides it is endlessly deep. And if you swerve from this path to one side or the other, you fall into that abyss. So you have to go along this path very carefully (Sahagun, translation by Lopez Austin).

A father giving advice to his son would refer to the wisdom of the ancestors, whose bones are in the soil.[6]

> . . . Indeed on a jagged edge we go, we live on earth. Here is down, over there is down. Wherever you go out of place to the side, wherever you take off to the side, there you will fall, there you will throw yourself over the precipice (Sahagun, translation by Burkhart).

'Tripping and stumbling, falling off precipices and into caves or torrents, appear over and over again in the sources as metaphors for moral aberration and its result' (Burkhart) Opposing poles should not be avoided completely, but must be balanced against each other. Walking on the ridged back of Cipactli (earth) implied the moral duty of careful balancing of the extremes to achieve a harmony of tensions. This shifting moral balance was expressed in people's careful and cautious pace on the narrow path that everyone had to trace on the corrugated skin of earth's surface. Undoubtedly, the mountainous geography of Mexico provided the ancient Nahuas with the metaphor of the earth as a giant iguana or alligator.

Aztec concepts of the divine

In Nahua religious thinking, the gods depended as much on humans as humans depended on the gods. All had a shared interest in the maintenance of the universe. Yet, the Aztec world was an animated world that had little place for the concept of an inert physical world ruled by a *deus ex machina*. Nahua deities were 'neither Aztec society writ large nor ethereal beings touching only tangentially on individual's lives' (Andrews and Hassig, Leon-Portilla). A permanent interaction characterized the relations between the Nahuas and their divinities. The sacred domain was not distant; it was a presence that suffused every element of nature, every daily activity, every ceremonial action and every physical being: flora and fauna, the sun, the moon and the stars, mountains, earth, water, fire were all divine presences. The Aztecs were so enmeshed in the 'supernatural' and the 'sacred' that the distinction between sacred and profane seems not to hold for them.

Sharing divine attributes with the god of duality were other forces, forces of nature that have been designated in popular thought as 'innumerable gods'. However, all these gods only embody the four powers that Ometeotl (two-god) has produced. They are the four elements: earth, air, fire, and water (Leon-Portilla). Each one of them was conceptualized as a dual female-male couple.

Divine earthly forces and earthly deities

The gods were not 'unique solutions', in that they did not have fixed unitary meanings (Hunt). One god could be conceived of as an aspect of another (Andrews and Hassig).

> The religious representation of earth in the symbolism of ancient Mesoamericans embodies some of their most complicated and diversified ideas. Since earth as a symbol complex was coded and transformed into practically all other mythic and ritual codes, it is impossible to produce a complete list (Hunt).

Earth, like the images of the gods, manifested a fundamental ambivalence. This ambivalence can be understood as the expression of the duality which pervaded all Mesoamerican constructs. The earth was both loving and destructive, both nurturing mother and carnivorous monster. Reflecting ideas both complicated and disquieting, the earth was often represented as a demonic figure (Hunt). We have seen that the mythic earth deity, Cipactli or Tlaltecuhtli, was a monster with a ridged back like an iguana, a giant frog or an alligator – the metaphor for the mountains and the creviced valleys of the earth's surface (Gonzalez Torres).

Tlazolli – dirt, mud, foul matter, soil

The concept of dirt cannot be separated from soil. To the soil we bequeathe our excrement; our bodies go back to the soil, and to the soil we let fall what is no longer useful.

In the Nahuatl language spoken by the Aztecs, there is a term that covers a whole range of impurities used in moral discourse to connote negativity. It is the word *tlazolli*, formed from the roots *tlalli* (earth, soil) and *zoli* (used, discarded). In its most literal meaning, it refers to something useless, used up, something that has lost its original order or structure and has been rendered 'loose and undifferentiated matter'. Broadly, it denotes any sort of dirt, chaff, straw, twigs, bits of hair or fibre, excrement, muck. What one sweeps up with a broom is *tlazolli* (Buckhart).

Yet, since most concepts were ambivalent, the word denoting 'filth' also had multiple favourable connotations, for maize grows from mud, from the body of the earth deity, and one linked oneself with the earth by eating cultivated foods like maize. Besides, all acts of *tlalticpacayotl* (earthliness, as we have seen, often understood as sexual activity) put people metaphorically into contact with *tlazolli*.

The *tlazolli* complex draws materials principally from the realms of excretion and decay to associate them, through the process of moral rhetoric, with less desirable activities. Manure used to fertilize crops is still called *tlazolli* (Burkhart). Therefore, some of these substances have a fertilizing, creative role.

Tlazolli was also the realm of the deity Tlazolteotl. She was the goddess who was responsible for sex and was associated with sexual transgressions.

Tlazolteotl, the goddess of 'filth'

Tlazolteotl, the deity associated with the sensuous, was the patroness of dust and filth, and of adulterers and promiscuous women. She had the power to provoke immoral activity as well as punish people for it. But she could also remove impurities. In that function she was called Tlaelcuani, 'eater of foul things', because she cleansed those who submitted to the indigenous confession rite by absorbing their impurities. This rite, as described in the *Florentine Codex*, was conducted by her diviners.

Tlazolteotl was closely related to the earth-deity complex (Sahagun 1989). Following Thelma Sullivan:

> (Tlazolteotl-Ixcuina), in her quadruple aspect as the four sisters, is a metaphor for the generative and regenerative cycle of life. Her fourfold character represents the growth and decline of things . . . She represents the Mother Goddess concept in its totality. This includes its negative as well as positive aspects (Sullivan).

Earth's womb

Caves were metaphorically referred to as the earth's womb. Similarly, the *temazcal*, 'sweat bath', shaped like a cave, was symbolic of the womb of Mother Earth (Sahagun).

Earthquakes were thought of in the same terms as uterine contractions: disorderly movement which could create but could also kill. The duality in the conception of a life-giving, life-destroying deity is evident here.

Tonantzin and Monantzin, 'Our Mother' and 'Your Mother' respectively, are titles of the Mother Goddess. Their name refers to the earth as the Great Womb. One incantation says, 'You (the seed) have been kept within Your/Our Mother' (Ruiz de Alarcon in Andrews and Hassig). Symbolism of the earth's interior, the mythical cave-house-uterus, still persists among contemporary Mayas in Zinacantan, Chiapas (Hunt).

Conclusions

Through this overview of the ancient Nahua concepts of the earth and divinity, especially as expressed in metaphors and attitudes concerning morality, we have come in contact with a distinct cosmovision. What bearing might it have on our contemporary ecological concerns?

Natural phenomena elicited awe in the Aztec mind. Physical beings were regarded as infused with the divine. Reciprocity with and understanding of other life forms is evident. This precludes abuse and exploitation of nature and natural resources.

Their entire belief system fostered and sustained a measured, non-exploitative use of the earth's resources. Aztec creation myths and stories did not give them the role of dominating nature, nor were they created as the species that ruled over all life forms. Rather, they were interconnected not only with 'nature' in the form of flora and fauna and with natural phenomena like wind and rain, but with the divinities that represented the entire natural domain.

This connectedness, however, could also prove fearful. The duality that pervades the Mesoamerican concept of the universe included both the positive and negative aspects of nature, the creative and the destructive, the nurturing and the annihilating forces. The metaphors for earth and nature were never romantic. We cannot conceive of the Nahuas, and this holds true for contemporary Indians, taking a stroll in nature. When they visit mountains and caves, it is to influence or placate the deities that live there. Because they have not lost their roots in nature, they still regard themselves as an integral part of earth. There is no sentimentality in their perception of the earth. Earth is a great nourishing deity and an unpredictable, fearsome monster: in all cases, it is necessary to move about on the earth with care.

In the moral domain, the *huehuetlatolli* speak often of the extreme care to be used walking on Cipactli's slippery back with an abyss on each side. Behaviour had to be such that balance was preserved – and this was a collective obligation. For Mesoamericans, appropriate behaviour while living one's life and enjoying the pleasures of earth was necessary to maintain the cosmic order.

Notes

1. Mesoamerican cultures extend over a broad geographical region which includes the Nahuas (of which the Aztecs were one group) and the Mayas among others. The terms Nahuas and Aztecs are used interchangeably.

2. The main sources are Fray Bernardino de Sahagun's interviews with Nahua elders as recorded in the *Florentine Codex (History of the Things of New Spain)* and also the work, attributed to Olmos. 'Historia de los Mexicanos por sus pinturas', in *Teogonia e Historia de los Mexicanos*, edited by A. Garibay.

3. These observations come from my own field data. This is corroborated by Eva Hunt (134–5).

4. Andres de Olmos, Gonzalez Torres, Sahagun, Lopez Austin.

5. See Marcos, 'Gender and Moral Precepts in Ancient Mexico: Sahagun's Texts', in *Concilium*, 1991/6, 60–74.

6. Bones symbolized for the Nahuas not only death but life and fertility. Life and death are in a dialectical (dual) relationship. Earth was both a tomb and an uterus. Lopez Austin reports the belief that semen originated in bone marrow (Lopez Austin 1988). Behaving morally entailed following the guidelines of the ancestors.

Bibliography

J. Richard Andrews and Ross Hassig, Introduction to *Treatise of the Heathen Superstitions that Today Live Among the Indians Native to this New Spain* (1629), by Hernando Ruiz de Alarcon, Norman, Oklahoma 1984

Louis M. Burkhart, *The Slippery Earth: Nahua-Christian Moral Dialogue in Sixteenth-Century Mexico*, Tucson, Arizona 1989

David Carrasco, *The Imagination of Matter: Religion and Ecology in Mesoamerican Traditions*, Oxford 1989

Yolotl Gonzalez Torres, *Diccionario de Mitologia y Religion de Mesoamerica*, Mexico 1991

Doris Heyden, 'Metaphors, Nahualtocaitl, and Other "Disguised" Terms Among the Aztecs', in Gary H. Gossen (ed.), *Symbol and Meaning Beyond the Closed Community: Essays in Mesoamerican Ideas*, Albany, NY 1986

Eva Hunt, *The Transformation of the Hummingbird: Cultural Roots of a Zinacantecan Mythical Poem*, Ithaca, NY 1977

Patrick Johansson, *La Palabra de los Aztecas*, Mexico 1993

T. J. Knab, 'Metaphors, Concepts, and Coherence in Aztec', in Gary H. Gossen (ed.), *Symbol and Meaning Beyond the Closed Community: Essays in Mesoamerican Ideas*, Albany, NY 1986

Miguel Leon-Portilla, *Aztec Thought and Culture*, Norman, Oklahoma 1990

Alfredo Lopez Austin, 'Cosmovision y Salud entre los Mexicas', in A. Lopez Austin and C. Viesca (eds.), *Historia general de la Medicina en Mexico*, I, Mexico 1984

——, *Cuerpo humano e Ideologia* (2 vols.), Mexico 1988

Sylvia Marcos, 'Gender and Moral Precepts in Ancient Mexico: Sahagun's Texts', *Concilium* 1991/6

Andres de Olmos, 'Historia de los Mexicanos por sus pinturas', in A. Garibay (ed.), *Teogania e Historia de los Mexicanos. Tres opusculos del siglo XVI*, Mexico 1973

H. Ruiz de Alarcón, 'Tratado de las Supersticiones de los Naturales de esta Nueva España', in Pedro Ponce, Pedro Sanchez Aquilar et al., *El Alma Encantada*, Mexico 1987

——, *Treatise on the Heathen Superstitions That Today Live Among the Indians Native to this New Spain* (1629), Norman, Okalahoma, 1984

Fray Bernardino de Sahagun, *Historia General de las Cosas de Nueva Espana*, Books I and II, introduction and paleography, etc., ed. A. Lopez Austin and J. Garcia Quintana, Mexico 1989

——, *Florentine Codex: General History of the Things of New Spain*, translation by A. Anderson and C. Dibble, Salt Lake City, 1969

Jacinto de la Serna, 'Manual de Ministros de Indios', *El Alma Encantada*, Mexico (1650) 1987

Rémi Siméon, *Diccionario de la lengua nahuatl o mexicano* (1885), Mexico 1988

Thelma Sullivan, 'Tlazolteotl-Ixcuina', in *The Art and Iconography of Late Post-Classical Central Mexico*, ed. Elizabeth Hill Boone, Washington, DC 1977

II · A Theological Reflection

The Cry of the Earth?

Biblical perspectives on ecology and violence

Christoph Uehlinger

> The earth is a battle, an unceasing battle: making it arable, planting, weeding, watering, until the harvest. And then you see your ripe field lying before you in the dew of the morning and you say to yourself, 'I, so and so, am the master of the dew,' and arrogance creeps into your heart. But the earth is like a wife: if one constantly mistreats her, she resists. I have seen how you have cut down the trees. Now the earth is naked and unprotected. It is the roots which establish friendship with the earth and support it: the mango trees, the oak woods, the mahogany trees, which give it rainwater for its great thirst and protect it from the midday heat with their shadow. That is the case, and it cannot be otherwise; otherwise the rain would erode the earth and the sun dry it up, and there would be nothing left but stony ground.
>
> I am speaking the truth. God has not abandoned human beings, but human beings have abandoned the earth and now are being punished – by drought, misery and desolation.'
>
> 'I don't want to hear any more,' said Delira, shaking her head. 'Your words sound like the truth, and perhaps the truth is a sin.'
>
> Jacques Romain, *Masters of the Dew* (1944)

These few sentences from the best-known work of the Haitian novelist Jacques Romain may remind theologians of the so-called 'First World' that for the majority of people the problem of ecology primarily presents itself in the framework of the everyday struggle for survival, as a problem of their own subsistence. Using the extreme example of the deforested hills of Haiti, they conjure up the wretchedness of poor farmers whose fields no longer bring forth any crops, because the earth is exhausted and has degenerated into a burnt and stony wilderness. Manuel, the hero of the novel, who comes home to the village from the city, and Delira, his old mother, interpret the wretchedness in different ways. Manuel rejects the view that the ecological catastrophe has been willed by God and rejects Delira's fatalistic attitude. Delira begins to suspect that her wretchedness and the wretchedness of the earth are two sides of the same coin. Delira's fear of the truth makes the scene seem both uncanny and realistic. Equally uncanny and realistic is Manuel's insistence that the poor themselves play a part in the destruction of the earth and are deeply entangled in the 'structures of sin', both as agents and as victims.

How can the cry of the poor and the cry of the earth be communicated in an organic theology of creation? This article seeks to contribute some biblical perspectives to the discussion. There can be no question of looking for answers to our present questions in scripture. Our re-reading can only seek to note attentively how some biblical writers have made problems at the interface of ecology and violence which arose in their own world a topic for discussion.[1]

I The land as a basis for life

Compared with the neighbouring regions by the Nile and the Euphrates, ancient Palestine was always quite a poor land. This was first of all because of its climate and its situation bordered by the sea and the wilderness, but also because of the marked internal divisions within the land, which at all times favoured a cultural and political regionalism. The land has no mineral resources other than deposits of copper and iron on the edge of the Arabah; these were known to the biblical authors (Deut. 8.9), but could only be worked to any great extent in the Byzantine period. Palestine was not a destination for the international trade between Egypt and Syria or Mesopotamia, or between Arabia and Rome, but always a not very attractive country to be passed through on the way.

How did the inhabitants of Palestine see ecological problems during the first millennium BCE? Primarily, and for the great majority of the population, they were a problem of subsistence. For the farming population of early Israel possession of land was *the* indispensable foundation of life – that can be taken for granted and is the presupposition for any biblically informed 'theology of the earth'.[2] Land, cultivated fields and gardens and some cattle, sheep or goats, were among the fundamental values of early Palestinian society – the basic necessities which enabled a free farmer to feed his family. For this purpose he needed not only the necessary economic resources, but also a minimal ecological security. In the rhythm of day and night, summer and winter, cold and heat, he recognized the working of God's blessing, which guaranteed that after a few months the sowing would be followed by a harvest (Gen. 8.22; cf. Ps. 65; 126.4–6). Periods of drought, damage, disease and the destruction caused by war constantly led him to experience the fragility of this equilibrium, which could be influenced only to a limited degree.

Numerous families on the land lived by a typical mixed economy of agriculture and the rearing of sheep and goats. They would concentrate more on one or the other, depending on the political and ecological situation. The degree to which they were aware of the sensitive balance between cultivated land and wilderness is shown by the many texts which speak of the curse of the encroachment of the wilderness or describe the rapidity with which thorns and thistles can grow over a neglected farm or vineyard. That this balance also included the relative security of human beings and herds from attacks by wild animals is emphasized in, for example Ex. 22.29: Yahweh is said not to have driven out the Canaanites in the very first year of the settlement, because otherwise the land would have become desolate and the wild animals would have gained the upper hand, to the detriment of the people. According to II Kings 17.24ff., the exiling of the population of northern Israel is said to have resulted in a plague of lions. The consciousness that in the battle against the wilderness and wild animals human beings were not *necessarily* stronger and the fact that in neighbouring woods or on the edge of the wilderness one could be attacked by a frightened panther, lion or bear (cf. Gen. 37.33; I Kings 13.24; II Kings 2.24; Amos 5.19) should be noted, if we are to have a proper understanding of the anthropology of Gen. 1.26–30 with its anthropology of dominion (see below, III.2).

The regulation that fields should lie fallow in the seventh year, which has pre-Israelite roots, is an expression of the awareness of what today we would call ecological balance. Exodus 23.10–12 specifies that this practice should be to the advantage of the poor and the beasts of the field. The commandment stands between regulations about the rights of the poor and the stranger, and the commandment to rest on the seventh day so that dependent workers (both human beings and animals) can have a breathing space. Thus social equilibrium and concern for animals are interwoven. In Deuteronomy the accent moves more markedly in a social direction, when the custom of leaving the fields fallow for a year becomes a social year of redemption (15.1–11). Conversely, the Holiness Code (Lev. 17–26) argues in an 'eco-theological' direction: the land itself is to keep sabbath for Yahweh; this is a time of regeneration in which it does not need to bring forth produce (25.2–7, 18–22). In the last resort the land belongs only to Yahweh (v. 23); when it keeps sabbath rest, it does so 'to the glory of

Yahweh'. Here Yahweh's privilege as owner of the land stands in the foreground. However, Yahweh's claim is not aimed at a privileged enrichment, which he does not need, but at a regeneration to which the land (like human beings and animals) has a right.

Leviticus 26.34f. and 43f. show how seriously Yahweh takes this right: Israel is to go into exile because it has despised Yahweh's precepts. From an anthropological perspective, the period of the exile represents only a period of devastation and catastrophe – this is the majority view in the First Testament. But here it is understood in positive terms as a period of rest for the land in which the land is to have restored to it the sabbaths which are its due and which have been abused by Israel (cf. II Chron. 36.21).

As was indicated earlier, Palestine was always a relatively poor land. So poor, that some people could make the accusation that this land 'devoured its inhabitants' (Num. 13.32; Ezek. 36.13f.). The authors of Num. 16 and Ezek. 36 defend themselves against this accusation and regard it as a slander. According to Num. 16, Moses, i.e. the Torah, also belongs to the land. If Israel observes the Torah, which requires life for all *and* respect for the land, then despite all the precariousness, it can also live in this land. Those who really dedicate themselves 'with all their hearts and all their minds and all their strength' (Deut. 6.5) to the realization of a society of brothers and sisters in solidarity will also be able to recognize a poor land as a 'good land' (Deut. 8).

II The cry of the earth

The language of the Hebrew Bible knows more than twenty verbs for crying and groaning, moaning and lamenting. Although some of them occur only very infrequently, this is a clear indication of the significance of distress and lament in the First Testament. Israel's history as the people of Yahweh began with the cry of distress of exploited foreign workers (Exod. 2.23f.; 3.7, 9; Deut. 26.7). This confession – which is not a historical but a religious statement – is meant to give the history of Israel a special significance. In it the cry of the poor will not remain unheard;

> If a stranger, a widow or an orphan cries out to me, I will surely hear their cry (Exod. 22.20, 22).

However, only once in the Bible is there any mention of the 'cry of the earth' (see below). This restraint is explicable in historical terms and is connected with the ideological process of 'de-divinizing' the world which took place from the fifth century BCE on. Jewish monotheism stands on the threshold leading from ancient Near Eastern mythology to the dedivinized thought of modern times. In the Bible the earth is only rarely personified in such a way that it appears as an autonomous subject. It is spoken of in a decidedly mythological way even more rarely (cf. Ps. 139.15; Job 1.21; Sir. 40.1 on the earth as 'mother', always in connection with the birth and/or burial of individuals, i.e. in the sphere of private piety).

1. Job's land cries out

The great apologia of Job, the ruined city patron (chs. 29–31), also comes to speak of his property, right at the end.

> If my land has cried out against me, and its furrows have wept together;
> if I have eaten its yield without payment, and caused the death of its owners;
> let thorns grow instead of wheat, and foul weeds instead of barley (Job 31.38–40).

The first sentence seems to presuppose that Job is speaking of farmland which belonged to him when he was still living with great riches (however, the personal pronoun is missing in both the Greek and the Syriac translations). But v. 39b speaks of other 'owners' (*ba'alim*). Who is meant? The majority of exegetes assume that Job is speaking of *former* owners of farmland which now belongs to him. In that case Job's hypothetical transgression would be that he had seized this land in a wicked way, by extortion or even murder (cf. I Kings 21),[3] and had cheated the former owners. It is also conceivable that former landowners who had been overwhelmed by debt have in the meantime become Job's unfree subjects, leaseholders or day-workers (v. 39a). Job would be being charged with having exploited the desperation of these people or even of having stirred them up in order to be able to extend his property. Verse 39a could be addressing the problem that Job had cheated his workers of their wages (Deut. 24.24f.; Lev. 19.13; cf. Jer. 22.13) – but that does not explain v. 39b (however, some commentators emend the text and read *po'alim*, 'workers', or something similar for *ba'alim*).

Elsewhere in the Hebrew Bible, the verb used for crying (*za'aq*) in v. 38a always has as its subject the direct victim of a transgression, usually a person against whom violence had been done (cf. Exod. 22.22; Deut. 22.23ff.; Job 16.18, etc.). That the fields are said to 'cry out' seems to presuppose that they themselves have also been violated or abused. One thinks most naturally of over-use of agricultural land, for example through failure to observe the regulations that it is to lie fallow (Exod. 23.10f.; Lev. 15.1ff.; see I.1 above); taken by itself, v. 38 is probably to be understood in this way. But v. 39 gives the crying and mourning of the land a wider meaning: it seems as if the earth itself is indignant about being misused by the unscrupulous landowner to exploit the workers who are dependent on it.

Given the topic of this issue of *Concilium*, we should note that the text regards the heedless exploitation of agricultural land and the exploitation of dependent workers as two facets of the same unscrupulousness, and reckons with a kind of 'ecological vengeance'[4] for both. However, the maltreatment of Job's farmland would not bring either comfort or justice to the exploited workers who had been robbed of their wages or their property. But what is at issue here is not a legal theory, an abstract principle of the reciprocity of transgression, consequence and punishment or reparation. Rather, here Job is considering specific charges of the kind that must have been made regularly against landowners of his status in fifth-century BCE Palestine. The three verses formulate in a way which, despite all the metaphor, is very realistic, something that can still be observed on the large estates in the twentieth century: over-use of the soil leads to a decline in yield, and heedless accumulation of land at the expense of the landless poor leads to the neglect of whole areas of farmland.

If my observations are correct, this is the only passage in the Bible which speaks explicitly of a 'cry of the earth'. So modern theological talk of the 'cry of the earth' can hardly take up the Bible directly. This evidence and the metaphor generally indicate one problem above all: the earth does not cry, because it cannot cry, although every day it is the victim of human and technological aggression, exploitation and violation.

2. *Ecological disaster and social sin*

The awareness that human intervention in the natural environment could produce irreversible destruction in the biosphere still lay far beyond the horizon of Palestinian farmers and the men of letters to whom we owe the Bible. However, what they knew from experience were failed harvests, fluctuations in yields, drought, ecological disaster as a consequence of wars – catastrophes, which resulted in famines and plagues and could force people to emigrate. Biblical texts often understand such disasters as the consequences of sin, a sin which could take many forms but which ultimately represented contempt for Yahweh, the Lord of the land. This is vividly described, for example, in Hos. 4.1–3.

Yahweh has a controversy with the people of the land.
There is no faithfulness or kindness,
and no knowledge of God in the land;

> there is swearing, lying, killing, stealing, and committing adultery;
> they break all bounds and murder follows murder.
> Therefore the land mourns, and all who dwell in it languish,
> and also the beasts of the field, and the birds of the air;
> and even the fish of the sea are taken away.

Social sin has catastrophic consequences in the ecological sphere: the earth fails to yield its produce, and its inhabitants, including animals, therefore perish. This represents no less than a reversal of the order of creation (cf. Gen. 1.28). Similarly, in Jer. 5.20–25, the 'house of Jacob' in Judah is accused of having destroyed the cosmic order by its sins, which is why no rain can fall (cf. Jer. 23.10ff.). The catastrophic consequences of the drought are depicted in the popular lament in Jer. 14.2–9, 19–22: the land becomes dry, the wells are empty and even the hind leaves her young in the lurch. Shock and bewilderment are widespread.

A specific experience lies behind such texts (cf. also Amos 8.4–8; Zeph. 1.2–3; Isa. 24.1–6; 33.7–9): in antiquity political, military and social conflicts usually also had effects in the ecological sphere. Remarks like those made in Hosea 4, that the lack of a knowledge of God, an absence of solidarity, political instability, mistrust and lies resulted in ecological devastation, are not just the expression of an obsolete 'mythical' and holistic view of reality. They reflect concrete historical experiences. Phenomena which were perceived as interconnected were also held together in faith and thought.

'This logic is at the same time both archaic and modern.'[5] A farmer from Haiti has no difficulty in understanding the history of dictatorships, following the book of Hosea, as political prostitution. And today all over the world we can see the degree to which wrong political and socio-economic developments can result in political disaster. In a world of highly specialized experts, who increasingly know more and more about less and less, synthetic thought has become difficult, but is needed more than ever as *critical* thought.

3. *'Holistic thinking' and ideology*

The link between guilt and the devastation of the land which is depicted in Hos. 4 and Jer. 5 seems to us today, who are fascinated by systemic or 'networked' thought, to be uncannily topical. However, we should not overlook the fact that it is permanently used by fatalistic and fundamentalistic theologies to form quite misleading links. The problem can be illustrated from the popular lament in Jer. 14; its particular drama lies in the fact that it fluctuates between the people's acknowledgment of guilt

(v. 7, 'Our iniquities testify against us . . . our backslidings are many') and its complete lack of comprehension (v. 9, 'Why should you be like a man confused, like a mighty man who cannot save?'). The people speaks of its own guilt, yet without being able to name it. The confession of guilt (cf. also v. 20) seems in any case to have been spoken only in despair, in order that God may finally turn the distress to good effect.

The explanation of an ecological disaster in terms of the guilt of the people is theologically legitimate when it leads to a better analysis of reality, to the identification of real social causes of ecological crises and thus also to a fight against the structures of sin. Where the network of ecological catastrophe and collective sin helps the ideological veiling of guilt and responsibility and encourages resignation, false prophets are at work.

4. A guilty land?

In Hosea 2 the land appears not only as the victim of sin but also as a guilty mother from whom Israel is to turn away and whom it is to accuse. Yahweh is even ready to mistreat this mother:

> Accuse your mother, accuse her –
> for she is not my wife,
> and I am not her husband.
> She is to put away her harlotry from her face,
> and her adultery from between her breasts;
> lest I strip her naked
> and make her as in the day she was born,
> and make her like a wilderness,
> and set her like a parched land,
> and slay her with thirst (Hos. 2.5).

The mother accused of adultery by her husband (Yahweh) is the land; prostitution is an image for the policy of alliances, trade and religion which drove the northern kingdom, Israel, into countless alliances and dependent relationships. What may seem to those involved as unavoidable, obvious and/or normal in real-political terms is criticized in an extremely polemical way by Hosea as going with strangers. Behind his criticism lies the religious conviction that Yahweh alone, and no one else, can bring good fortune to the land.

Behind the metaphors of marriage and adultery stands a 'Canaanite' religious tradition, evidently also known in pre-exilic Israel, which imagined the storm god (rain from heaven) and the vegetation goddess (the flourishing of the earth) as a couple and depicted their interaction as sexual intercourse. There are two modifications in Hosea's use of this metaphor:

the land is not described directly as a goddess but as a mother, and the relationship of the couple is linked with the legal notion of marriage and adultery. The patriarchal image of God associated with this and the sexism expressed in the metaphors of harlotry and mistreatment compel us to read this text particularly critically today. That fact is not altered even by the conclusion of Hos. 2, according to which the husband Yahweh in the end does not mistreat his wife but wants to resort to other means. He wants to beguile his wife again, and an erotic relationship is to replace that of mastery and possession (2.16–18).

Can the land, the mother, be punished for transgressions of its children, the people, or the social elite? This question is not just a modern one. Even the most obvious re-reading of Hos. 2 within the Bible, namely Ezek. 16, corrects the metaphor and puts the states or peoples of Israel and Judah in the role of wives who prostitute themselves (Ezek. 16). The difference between the two signalizes progress in theological knowledge which recalls the problem of Num. 16: it is not the land which is bad, but the way in which people treat it. As often in the Bible regression (even more sexism in the metaphor of prostitution) is bound up with progress (in the question of responsibility).

III Dominion of the earth? A critical re-reading

The heavens are the Lord's heavens,
but the earth he has given to the sons of men (Ps. 115.16)

Even if some recent historical investigations tend more to relativize the significance of Gen. 1.26–30 for the development of the modern Western understanding of nature and for the ideological justification of the technical subjection and exploitation of 'nature',[6] the doctrine of the dominion over the earth by human beings nevertheless lays a heavy burden on the Christian theology of creation and on Christian theology. Anyone concerned with the topic of ecology and violence in the Bible cannot avoid grappling with Gen. 1.26–30.

1. The need for re-contextualizing

A look at the more recent exegetical literature on Gen. 1 and dominion over the earth quickly makes clear how such exegesis is a child of its time and in what social context it is practised. In 1934 the exegete Benno Jacob could put forward the view that with Gen. 1.28 'unlimited domination over the planet earth is given to human beings, and therefore no work on it, i.e. tunnelling through or removing mountains, drying up and diverting rivers and the like, can be described as violence which is contrary to God. Human beings can sin only through immoral action on the earth, as a result of which they desecrate it and which it senses' (cf. Gen. 6.13). However – and this is the significance of the qualification –, they cannot sin against the earth.[7] This position could be maintained through the period of the Second World War and the optimism of post-war progress, down to the early 1970s. After the appearance of the first report of the 'Club of Rome' (*The Limits of Growth*, 1972), the exegetical pendulum swung a long way in the opposite direction. Distinguished exegetes now thought that they could recognize in Gen. 1.26–30 the commissioning of a steward to deal carefully with the goods entrusted to him in the 'house of life'. We can recognize clearly from a distance of between one and two decades that often an unexpressed apologetic interest was at work here, but we may not resort to cheap criticism of it. For at any rate these works introduced a re-orientation of method to which even the most recent contributions on the subject are still indebted, and which we can describe as re-contextualization on several levels:

- More marked attention to the text and a semantic analysis of the key Hebrew words (*rdh* in Gen. 1.26, 38 and *kbš* in Gen. 1.28);
- An analysis and interpretation of Gen. 1.26–30 in the framework of its literary context, namely (a) the creation narrative of Gen. 1.1–2.3 (,4a) which comes to a climax in the hallowing of the seventh day and God's resting; (b) the Priestly creation story (above all in terms of the analogies and differences from the blessing, including the covenant with Noah, which is given in Gen. 9 after the flood); (c) the Priestly Writing as a whole, the culmination of which lies in the appearance of the presence of God in the tabernacle in the midst of the people (Exod. 24, the sanctuary as a model of the creation).
- Increased questioning of the pragmatics of the Priestly creation narrative and the reason for it in the historical context of its original authors and audience.
- Coupled with this, a move from a dialectical contrasting of Gen. 1 with ancient Near Eastern creation myths to the question of models of thought which can make Gen. 1 comprehensible as a text from the ancient Near East.
- A relativizing of Gen. 1 as one biblical text on creation among many, contrasting the anthropocentric perspective with less anthropocentric texts (Ps. 104) or texts which are not anthropocentric at all (Job 38–39).

To this list of re-contextualizings we should add yet one more, the need to take into account the socio-cultural and historical context of contemporary exegetes. From the limited perspective of a European exegete, it seems to be predominantly Christian, German-speaking exegetes who have taken part in more recent exegetical discussion of Gen. 1.26ff.[8] The widespread lack of specifically feminist exegetical works on the topic in European discussion is also striking. Feminist theologians often rightly point to the connection between the destruction of the environment and the oppression of women,[9] but tend to concentrate on a correction of the traditional patterns of interpretation of Gen. 2–3.[10]

2. The primacy of the blessing and the right to living space and subsistence

As far as I can see, nowadays there is widespread unanimity that Gen. 1 does not offer *carte blanche* for the exploitation and plunder of 'nature'. If God's creative action in Gen. 1 introduces ordinances into chaos which make life possible, the task of humanity that is created after the model of God cannot lie in the destruction of this life but only in its preservation and cultivation. It is by exercising this function of order that humanity realizes the 'image of God' which it bears. This denotes a function (being representatives of God) which the author of Gen. 1 is convinced can be exercised only by human beings, since only they are characterized by a personal will which makes them capable of being God's partner or God's representatives. God gives humankind a special power to exercise this unction, which also implies a special responsibility:

The key verbs *rdh* in Gen. 1.26–28 and *kbš* in Gen. 1.28 cannot be completely 'pacified'; they imply the possibility, indeed the necessity, of exercising authoritative control. They are terms which are rooted in the kingship ideology of the ancient Near East and in popular law. But – to put it bluntly, in accordance with the self-understanding of kingship ideology – the task of an ancient Near Eastern king was to exercise power and to limit violence.

A glance at the immediate context is all the more significant: *kbš* means to 'tread on the earth' in the sense of 'subjecting it, claiming it and making it serviceable'. Here the cultivation of the earth by agriculture is also envisaged, but not exclusively. Rather, humankind in general is authorized to claim, open up and use the earth as the living space which has been granted it. The significance of the right to rule in no way involves the destruction of living space. Human beings need the earth in order to be able to live.[11]

However, the earth must literally be colonized as living space, and here humankind has to compete with the land animals. In vv. 25ff. these are not blessed by God; the blessing to multiply remains reserved for human beings: a clear indication that the author regards human beings as under greater threat than animals. *rdh* means 'rule', not in the sense of a special measure or function like 'tame, lead, pasture', etc., but focussing on a universal function of human beings to bring order among other living beings.[12] This implies supremacy,[13] among other things for keeping flocks and domestication – but not the right to kill, as is shown by the fact that according to Gen. 1.29f., both human beings and animals are vegetarian. Only in Gen. 9.1–5 will this utopia be withdrawn and will human beings be authorized to kill animals.

In Gen. 1 the capacity of human beings to rule which is given with the image of God is completely orientated on the non-human environment (earth and animals). Humankind has power only over them, and only so that it is not a prey to their violence. This power serves exclusively to safeguard living space and food and thus is directed towards subsistence. The fact that the high-point of creation lies in the sabbath day rest on which the earth (with its own blessing) can flourish once again without humanity having to exercise its rule indicates that the creator God is concerned with limiting power to what is necessary for survival.

Genesis 1 (in contrast to Genesis 9) does not regulate any form of violence between human beings. The text sees humankind as a whole and as man and woman appointed to be representative of God and thus postulates a fundamental equality of all human beings and both sexes where the right of subsistence and dominion of the earth is concerned. But it also follows that ecological power is not covered by Genesis 1 wherever it presupposes social (political, social or economic) violence.

Genesis 1 is a utopian text. The authors were already aware that this cosmic ideal order did not correspond to reality, which is shaped not only by the war between human beings and animals but also by social violence among human beings. What is *not* regulated in Genesis 1 – violence among animals and among human beings – subsequently develops in an anarchical and life-threatening way and finally leads to the Flood (Gen. 6.11f.). What subsists after the Flood is a less perfect but more realistic order in which account is taken of the existence of violence, but this is regulated, and at the same time it is inculcated into human beings that God's covenant applies to the whole of creation (Gen. 9).

3. The limits of the authorization to rule in Genesis 1

As modern readers, concerned with contextualization, we must refuse to generalize over-hastily about the validity of Genesis 1 or its ability to give us orientation today. We must not speak unthinkingly of 'human beings' or 'humankind' generally. It makes a great difference whether one applies the command to cultivate and the authority to rule to a landless *campesino*, a landscape gardener in a European country of the chairman of a multinational corporation involved in building dams. Genesis 1.26–30 legitimates the exercise of human domination over the earth and animals because in the view of the author such rule could not be taken for granted and needed special legitimation. The conception of the domination of the earth in the Priestly Writing is also based on an unconditional respect for the whole of creation. Gen. 1.28f. stands under the primacy of the blessing. Anyone who no longer needs this blessing or thinks it necessary to be emancipated from it cannot appeal to the authorization to exercise rule.

The transference of a royal role to the whole of humankind, both men and women, under the primacy of the blessing, makes Gen. 1 seem to be a strong, challenging, but also very problematical text in the modern, global debate on ecology. In my view, its strength lies in its potential to offer criticism of rule and the limitation of the dominion of the earth to the right of subsistence. Genesis 1 can remind us that questions need to be asked about technological progress in the utilization of 'nature': does it respect the rights of all humankind and all human beings, men and women, to living space and basic subsistence, or on the contrary does it increase existing concentrations of power and consolidate structures of exploitation?

The problems associated with Gen. 1 (cf. also Ps. 8) lie in the undisputable, basic anthropocentrism which makes the earth primarily appear as an object of human exploitation, and makes animals competitors and subjects both to human control. Here Gen. 1 is certainly opposed to Ps. 104, which depicts a harmonious cohabitation of animals and human beings, in which each species has been given its due living space and necessary nourishment by the creator God. For the author of Gen. 1 the right of human beings to subsist on the earth and alongside the animals is not guaranteed, but needs mythical legitimation; Psalm 104 presupposes it and suppresses the competition between human beings and animals which in fact exists. Neither of the two texts is 'realistic'; both texts are utopian, and the difference between them is primarily determined by the genre.

The deep ambivalence of Gen. 1 is ultimately connected with the paradigm of responsibility given with the metaphor of kingship: talk of the royal man can easily turn into an irresponsible idealism, kingly rule into tyranny. Therefore questions always need to be asked about the doctrine of having dominion over the earth, about which human beings are in fact in a position to exercise kingly rule, which human beings are systematically and structurally excluded from this power, and which human beings, institutions and states misuse their social and ecological supremacy irresponsibly and without paying heed to present or future victims.

4. The cry of joy from God's plundered garden

The Hebrew Bible knows a number of cases in which the possibility of the royal man to control the earth and the world of plants and animals has been misused, and it clearly brands the perpetrators as exploiters and oppressors. The following lines come from the taunt song on an anonymous 'king of Babylon' who has fallen and been robbed of all his glory (v. 4a). Behind this king is certainly the historical figure of the Assyrian king Sargon II, who perished in a battle in 705 BCE. The inglorious fate of the ruler of the world was understood as divine punishment not only in Assyria but also in all the provinces.

> The whole earth is at rest and quiet;
> they break forth into singing.
> The cypresses rejoice at you,
> the cedars of Lebanon, saying,
> 'Since you were laid low,
> no hewer comes up against us' (Isa. 14.7–8).

The fallen king had not only made individuals and nations bow under his yoke, but he had also attacked the forests of Lebanon. To whom did these woods belong? In the first millennium the kings of Assyria (cf. Isa. 37.24 = II Kings 19.23), Babylon and Persia had claimed the right to exploit Lebanon. However, an Egyptian papyrus from the eleventh century BCE claims that the Lebanon and its trees belonged to Amon, the supreme god of the Egyptian kingdom. Still, an emissary from the king of Egypt sent to Byblos to get wood for building came up against vigorous resistance from the local king. Because of its almost inpenetrably dense forests the Lebanon was regarded by the inhabitants of the Levant as 'God's garden' (Ezek. 38.8f.). According to Ps. 104.16, Yahweh planted the cedars of Lebanon with his own hands (cf. Ps. 80.11). As the psalm may have originated in Phoenicia, Yahweh probably succeeded to this role of 'Lord (*ba'al*) of Lebanon'.

Thus a divine accusation against the empires was a repeated feature of the history of Lebanon. In Isaiah 14 the liberated trees of Lebanon rejoice at the death of the tyrant. The cry of the trees is echoed in the cry of the poor. Their jubilation is an expression of the protest which the bordering lands, rich in raw material but politically dependent, have constantly made against exploitative central powers and their local minions when these have attacked even the holiest of places out of purely economic interest and without respect for the cultural traditions and values of the 'subjects'.

Today such tyranny is being unleashed with a brutality increased a millionfold on the rain forests of the Amazon and South-East Asia. Thus under the conditions of the present disorder of the world economy, Isaiah 14, too, has only the hopeful ring of utopia.

Conclusion: Against the temptation of false alternatives

A theology of creation which even today still begins in a one-sided way with dominion over the earth and the paradigm of responsibility connected with it indulges in an illusory and ultimately irresponsible idealism. Modern technology has long since advanced into the experimental sphere in which human beings can no longer disregard the consequence of their interventions in the 'natural' order. In this situation a responsible way of dealing with the 'natural' environment can only be appealed for; in fact the demand can no longer be fulfilled. Even those who today want to devote themselves wholly to the protection of nature for the sake of human survival are not immune to contributing to its destruction.

In view of this situation, a contextually appropriate theology of creation in the countries of the North needs, first, a criticism of ecological tyranny. Secondly, it urgently and clearly needs non-anthropocentric perspectives which could lead to an ethic of self-limitation or avoidance. Within the framework of this article it has not been possible for me to go more closely into what the Bible offers in this connection. Psalm 104 and Job 38–39 are the best texts to serve as guidelines for a non-anthropocentric theology of creation.[14] The theocentricity of both texts can be connected with a relational theology which seeks to go behind the subject-object dichotomy of modern Europe.

A criticism of ecological tyranny is also urgently needed in the lands of the South. However, here the need for discussion seems to me, for various reasons, to be somewhat different. First, quite often there is a fund of indigenous non-anthropocentric traditions which understand human beings as part of a web of creation. These traditions can serve better as the foundation of an inculturated theology of creation which is contextually more appropriate and ecologically more viable than many biblical texts. They make it easy to gain a critical distance from the one-sidedly anthropomorphic thought models of the Bible. On the other hand, liberation theologians rightly continue to remain at a critical distance even from the more recent sketches of a creation theology in the North, when these propagate a departure from anthropocentric thought-models without at the same time arguing resolutely for the right of the poor to life and without making it clear that they are primarily referring to the imperialistic and patriarchal misuse of anthropocentrism. It is necessary constantly to put the question to non-anthropocentric schemes whether – either in theory or in practice – in marginalizing 'human beings' they do not also marginalize the poor. However, the basic right of the poor to living space and subsistence is an integral part of the creation which needs to be preserved.

On the bald slopes of Haiti and in many other regions of the world the earth is showing its resistance to reckless exploitation. The earth is defending itself, but it cannot cry aloud. How can those who will not hear the cry of the poor detect the cry of the earth?

Translated by John Bowden

Notes

1. The remarks which follow develop earlier comments written for European readers, cf. C. Uehlinger, 'Vom *dominium terrae* zu einem Ethos der Selbstbeschränkung? Alttestamentliche Einsprüche gegen einen tyrannischen Umgang mit der Schöpfung', *Bibel und Liturgie* 64, 1991, 59–74. Here I am attempting to think more markedly from the perspective of the poor. This has led to some shifts in emphasis. I am grateful to Detlef Hecking for his careful reading of the first draft of this article.

2. M. de Barros Souza and J. L. Caravias, *Teologia de la Terra*, Madrid 1987.

3. Naboth's field does not cry to heaven. Therefore exegetes often refer to Gen. 4.10f. as an explanation of Job 31.38. However, there it is the *blood* of the murdered Abel, i.e. his power of life, which cries out to God; by contrast the soil did not show solidarity with the murdered man, since it was ready to swallow up his blood (v. 11, cf. Job 16.18!; I Enoch 7.6; 87.1 differs).

4. L. Alonso Schökel and J. Sicre Dias, *Job. Comentario teológico y literaria*, Nueva Biblia Española, Madrid 1983, 447.

5. J. Ebach, 'Schöpfung in der hebräischen Bibel', in G. Altner (ed.), *Ökologische Theologie. Perspektive zur Orientierung*, Stuttgart 1989, 98–129 (on Hos. 4, esp. 100ff.).

6. Cf. J. Cohen, *'Be Fertile and Increase. Fill the Earth and Master It'. The Ancient and Medieval Career of a Biblical Text*, Ithaca, NY and London 1989.

7. *Das Erste Buch der Tora. Genesis*, Berlin 1934, 61.

8. J. Ebach, O. Keel, B. Janowski, J. Jeremias, K. Koch, N. Lohfink, H. D. Preuss, O. H. Steck, P. Weimar, E. Zenger, etc. Since the literature is so extensive I shall mention only the (provisional) culmination of it: U. Rüterswörden, *Dominium terrae. Studien zur Genese einer alttestamentlichen Vorstellung*, BZAW 215, Berlin 1993 (with bibliography). In many countries the discussion seems to have taken place more markedly at a pastoral level. But cf. Association Catholique Française pour l'étude de la Bible, *La création dans l'Orient ancien*, Lectio divina 127, Paris 1987; P. Gibert, 'Entre l'idée de la création et le récit biblique – Un point sur la question', *Recherches de science religieuse* 81, 1993, 519–38.

9. Cf. e.g. Anne Primavesi, *From Apocalypse to Genesis: Ecology, Feminism and Christianity*, Tunbridge Wells and Minneapolis 1991; Rosemary Radford Ruether, *Gaia and God. An Eco-Feminist Theology of Earth Healing*, New York and London 1992.

10. But cf. Luise Schottroff, 'The Creation Narrative. Genesis 1.1–2.4a', in Ahtalya Brenner (ed.), *A Feminist Companion to Genesis*, The Feminist Companion to the Bible 2, Sheffield 1993, 24–38.

11. The expectation of the exiled or returned Judaeans that they would again be able to take possession of their land (described in Num. 14.7 by Joshua as a 'good, *very* good land,' echoing Gen. 1) is set against the background of the authorization to dominate (cf. *kbš* with reference to the settlement in Num. 32.22, 29; Josh. 18.1).

12. Cf. Rüterswörden, *Dominium Terrae* (n. 8), 91ff., 105ff.; B. Janowski, 'Herrschaft über die Tiere', in *Biblische Theologie und gesellschaftlicher Wandel. FS N. Lohfink*, Freiburg im Breisgau 1993, 183–98.

13. *rdh* denotes varieties of kingly supremacy over foreign peoples (e.g. I Kings 5.4; Ps. 72.8; 110.2; Isa. 14.6; 41.2 – never the rule that a king exercises over his own people!), and Lev. 26.17 shows that to (post-)exilic Judaeans, being under such foreign control seemed to be a curse (cf. Neh. 9.28, and in connection with forced labour, I Kings 5.30; 9.23).

14. Cf. O. Keel, *Yahweh's Entgegnung an Ijob*, FRLANT 121, Göttingen 1978; cf. n. 1.

Spirituality of the Earth

Julia Esquivel Velásquez

Introduction

From the Central American viewpoint, it seems impossible to stop the process of turning the isthmus into a desert. This is not caused by ignorance or lack of studies or programmes, but by lack of responsibility, decision and political will. Mere rational conviction suffers from impotence because the earth's growing devastation is directly linked to the attacks, which still continue, on the indigenous and peasant populations, causing their increasing impoverishment and dependency. This perspective is also valid at planetary level and is particularly acute in rural areas all over the Third World.

The biologist Mary Mersky told me that she was convinced that without a change, a deep conversion on the part of governments and all of us, it will be impossible to recover the power to respect the earth's life and cooperate in its recovery, or at least defend what is left to us of nature. We Christians learned and believed that 'man is the king of creation'. This error led us to feel we were above all creation. Arrogantly we believed that we were masters.

Men with economic and military power have conquered, sacked, abused, sold the earth and treated it and all it produces and contains as simply for their use. They have sacrificed its wealth and beauty to produce capital and goods manufactured in their factories, in order to satisfy their lust to possess and to dominate. This behaviour demonstrates their spiritual emptiness, pride and vanity.

Ignorant of ecological laws, they have introduced chaos into the natural order. Where there were wonderful jungles, they have destroyed the vegetation to introduce enormous herds of livestock. They have deflected the course of rivers, burned forests and introduced species alien to the environment in places originally destined by nature for a different kind of fauna and flora.

And as if this were not enough, in 1984 the arsenals of the great powers included nuclear warheads whose accumulated power represented approximately one million two hundred times the power of the bomb that destroyed Hiroshima. Sometimes this power is translated as corresponding to four tons of explosive for every human being on earth.[1] The economic and military system constructed by Western human beings feeds insatiably on the constant destruction of the peoples of the South and of nature. It displays symptoms of spiritual emptiness and a destructive fury.

Human beings are creatures among other creatures

The indispensable condition for a true conversion is to put away pride, and honestly admit that we are mistaken. We are putting the earth and all life, including our own, in grave danger and we must humbly recognize that we are merely creatures on this earth.

In the Chouaqui version of the Bible, Adam in Genesis is called 'Le Glébeux' ('man made of earth, dust'). As dust, we are part of nature, formed from and dependent on the same primary material from which all creation is made.

From Einstein to our own days, physics and the discoveries of the new biology and ecology confirm that: 'The whole earth is a single cell, and we are all simply symbiotic particles, related to one another. There can be no "us" and "them". The global politics that flows (should flow) from this vision is truly *a bios* and *a logos*.'[2]

'Nature recycles its materials again and again without generating any kind of waste.' Dieter Teufel, of the Heidelberg Umwelt und Prognose Institut, has calculated that 'all the carbon there is in our bodies, our food, the carbon dioxide in the atmosphere and limestone rocks, has already formed part of other organisms six hundred times in the process of life production'. In the body of each one of us, there are about half a billion carbon atoms, which were part of the organism of persons living two thousand years ago, for example Jesus Christ. Likewise, according to Teufel's models, all 'the nitrogen there is on earth has already formed part of the organism of living beings and been eliminated from them approximately 800 times; the sulphur 300 times; the phosporus 8000 times; the potassium 2000 times', etc. Thus nature is the cleanest, most efficient, astonishing and instructive factory imaginable, an example which humans must follow if we want to survive.[3]

The whole creation's source of life is one and the same. According to the new physicists, this source is intelligent and communicates in energy impulses throughout all created things. Without it, it would be impossible for photosynthesis to happen, for our hearts to beat, or for any of the biological phenomena of living beings that we can observe, or the invisible energy movements we find in the subatomic world, to happen.

The new science shows us that we are related very closely not only to other creatures, but also to all things. In her book *Gaia and God*, Rosemary Radford Ruether says: 'If we tried to experience this relatedness and to keep it present in our awareness, an *intense spirituality* would flow from it.'

'Dust thou art and to dust thou shalt return' is precisely true. It reminds us that the generation and sustaining of life is one single process. We are inter-connected creatures, who need one another.[4] Another witness calling us to this humility is the relationship with nature of the original peoples of Abya Yala (the continent of America today).

A letter dated 1854, from Chief Seattle of the Duamich League of North Eastern territories, afterwards called the state of Washington, addressed to Franklin Pearce in reply to the proposal to 'sell their lands', is an example of this prophetic wisdom.

I can only quote briefly from it. The white man 'treats his mother the earth and his brother the sky as things to be bought and sold. His appetite will gobble up the earth and leave behind merely a desert . . . ' 'What is happening to the earth will happen to the earth's children. Man cannot control the web of life. He is only a thread in this tapestry. What he does to the tapestry he will do to himself . . . ' 'The air is precious to the red-skinned people because all things share the same breath: animals, trees and humans . . . ' 'The sap circulating in the trees carries the memory of the red-skinned people . . . ' 'Our God is the same God. You may think now that he belongs to you, just as you wish to possess our land; but it is not possible. He is the God of all human beings, and his mercy is equal towards the red-skinned and the white. This earth is precious to him, and violating it is despising its Creator . . . ' 'My words are fixed as the stars.'

Western Christian creation doctrine set up an abyss between 'the Christian' as a human being and all others. The Christian was above all others and even his own being was split into matter and spirit. Consequently during the Conquest Christians proceeded to baptize masses of Indians and then to kill them, with the justification that they were saving their souls. This same justification, of conquest in order to evangelize, betrays the contradiction that led to genocide.

Another example of this attitude which is in accord with the vision of the new science can be found in the Maya Tojolabal language. After his experience of living with the Tojolabales for twenty years, Carlos Lenkersdorf writes in his book *Lengua y Cosmovisión mayas en Chiapas* that in the grammatical construction of the Tojolabal language, the subject-object relationship does not exist, because it is not a language with an accusative character like the European languages. Existential and agent subjects exist in a dynamic, complementary and reciprocal relationship, which is impossible to translate into Spanish. The determining factor is the presence of 'multiple and qualitatively different subjects', through which events occur. The name we might give to this peculiarity would be inter-subjectivity. This excludes objects and, as it were, raises all to the level of subjects. According to Lenkersdorf, this linguistic structure corresponds to the Tojolabal way of relating among human beings and between all things. For them there is no dead matter: all things have a heart. Like other pre-Columbian peoples, they live immersed in a cosmic community: 'we all live and share life with everything in the cosmos.'

The Tojolabal language is another window into the Mayan soul. Here again we find that humble and respectful attitude to creation peculiar to the original peoples, which agrees with the vision of the new physics and biology. They have a way of life expressing a spirituality unknown to Christians, which we must try to understand. We must recognize that we are going the wrong way, in a direction leading us to destruction, and that 'there is no end to the chain of influences which are the consequence of my decision'.[5]

At individual, family, national and global level, each decision sets in motion an endless chain of influences either for life, health and the balance of nature or for the lessening of life, unbalance and chaos, in other words, death. The death of 40,000 children every day from poverty is caused by decisions taken at supra-national level. These affect the weakest peoples and sectors, when they are converted into economic plans the aim of which is the production of capital, arms and unnecessary things.

The destruction of communities of aborigines and in the Amazon jungle in Brazil is caused by the creation of 'artificial needs' by big companies, whose sole interest is in accumulating more capital in an endless cycle, which is like a whirlwind of death for human beings and nature.

A real conversion means changing our awareness, attitude, decisions and will, so that there is a real change in our way of life:

> Therefore, brothers and sisters, I pray you by the mercies of God, to present your bodies as a living sacrifice, holy and acceptable to God, which is your spiritual worship. Do not be conformed to this world but

be transformed by the renewal of your mind, that you may prove what is the will of God, what is good and acceptable and perfect (Rom. 12.1–2).

Changing: experiencing grace, gratitude and free giving

Perhaps only we human beings have the power to become aware of ourselves. If we were to heal ourselves, we could open our eyes and see who we really are. This sight would make us realize the miracle of life and the infinite wealth surrounding us, and give us insight into the enormous possibilities of being and of the whole creation. We can only receive this revelation through grace and compassion. What we are and have received has all been freely given. In Chief Seattle's letter there is infinite gratitude and respect for all that exists and profound astonishment at the white man's craziness and insensitivity.

Jesus tells us: 'Freely you have received, freely give.' Here it is not a case of giving, because we are not owners; we are receivers and beneficiaries like all creatures. It means gratefully receiving what we have been given and gratefully sharing it. God, the source of life, has offered us the possibility of becoming a reflection of his own being (like him), without expecting anything more or anything less from us than love.

This pristine vision is muddied by the selfishness of human beings, who instead of gratefully receiving, take over, dominate, commercialize and convert God's gifts into the golden calf, capital. Believing they are the creator, they have set their hearts on their 'possessions'. Jesus says of these people: 'It is easier for a camel to go through a needle's eye than for a rich man to enter the kingdom of heaven.' This spirit of gratitude and free giving nurtures the conversion necessary to save ourselves as a species, avoid the destruction of the planet, and consciously contribute to our own renewal and that of all things.

With new eyes and heart, all things will be made new, as they really are each morning, fresh from God's hands. This was the prayer of Chief Seattle, the spirit of St Francis of Assisi's *Canticle of the Creatures*, and is the key to Jesus' words: 'Is not life more than food, and the body more than clothing?' (Matt. 6.26) This is our own heart's desire. Encountering moment by moment life's abounding grace, we become aware of the joy of heaven and earth singing God's glory. This change of mind restores the damaged sensitivity of our whole being, so that we grasp the messages of wisdom emitted from one day to the next, pierce night's stillness and decipher the mysteries transmitted night after night (Ps. 19.2–3).

Grace, gratitude and free giving are resonances of love, notes in a single tune, compassion. Grace rescues us from the most subtle baseness, that of clinging to things and getting stuck to them. It leads us to give freely, to share. A true conversion leads us away from selfishness towards communion; from competition to co-operation; from plundering to giving freely; from covetousness to respect in our relationship with all creatures great and small.

Changing: healing and growing in order to assume our responsibility

Only those who know they are loved are capable of loving themselves and others. Beginning to love means beginning to heal. In the process, every natural happening recovers a deeper sense: rain which refreshes the earth, the air which is the breath of life; the warm kiss of the sun and fire, its child, our friend.

Gratitude makes us members of a great fellowship in which we become consciously responsible for our own inner development and a revolution in our relations with the earth, all creatures, all things, nourished like us by our common mother.

Our lives which have so often been turned in upon an isolated and fearful ego have not known true communion even with ourselves, let alone with others, creation, or the Creator. Healing means spreading our tent-pegs infinitely wider, transcending our individual ego, our own little family, country, religious circle, outwards to embrace the great family of the earth and the cosmos. True love is inclusive.

If human beings, particularly in the West, have not been able to have sane relations with their fellows and the planet, it is because they are sick. Nevertheless, our human potential can reach out for something better, because we are part of the intelligent miracle which creates and sustains life. Our deepest desire is to grow and share in our own process of evolution.

We Christians know that the Spirit which dwells in us wants us to grow. What we are now discovering with new awareness and new eyes is that all our seeking, the thirst in us for something greater, is the expression of that Spirit who constantly presses us on. From a wider viewpoint it is a call to co-operate and share in the growth and transformation of all created things.

> Yet among the mature we do impart wisdom, although it is not a wisdom of this age or of the rulers of this age, who are doomed to pass away. But we impart a secret and hidden wisdom of God, which God decreed before the ages for our glorification. None of the rulers of this age

> understood this; for if they had, they would not have crucified the Lord of glory. But as it is written,
> 'What no eye has seen, nor ear heard,
> nor heart of man conceived,
> what God has prepared for those who love him,'
> God has revealed to us through the Spirit. For the Spirit searches everything, even the depths of God. For what person knows a man's thoughts except the spirit of the man which is in him? So also no one comprehends the thoughts of God, except the Spirit of God. Now we have received not the Spirit of this world, but the Spirit which is from God, that we might understand the gifts bestowed on us by God (I Cor. 2.6–12).

Therefore our groans are united with all creation's (Rom. 8.18–30), even though we have become deafened through conforming to this system, which has exchanged spiritual growth for the thirst to possess and cling on to things. But this growth, which means understanding and responding to the Spirit's purpose, enables us to see why the Tojolabales believe that all things have a heart. For it is essential both for them and us to live in harmony with things, rather than dominating them.

Dr Derek Chopra says in his book *Cuerpo sin edad, mentes sin tiempo*: 'Although we often identify love with grasping and possession, there is a profound truth here: losing the power of detachment means losing the power to love.' Love is inclusive and expansive. It does not cease at the boundaries of our skin or our senses, or the walls of my house, or the frontiers of my country. If we cultivate love it expands infinitely.

People with many possessions lose their soul by clinging on to them. They are imprisoned or possessed by what they have accumulated. They are not free. They suffer an emptiness. They are ignorant and afraid of the meaning of fellowship, respect, gratitude and free giving. They have lost sight and hearing and heart. They are mortally sick.

In this context Einstein says: 'Our task must be to free ourselves from this prison, broadening our circle of compassion to embrace every living creature and the whole of nature in all its beauty.'[6] Love heals us, driving us out of a narrow sectarianism towards ever deeper and more inclusive communion, transcending frontiers in order to appreciate the wealth of diversity and glimpse a planetary home, in which each people and culture and religion, enthusiastic for the welfare of the world, asks itself: How can I contribute to humanity? What fruits can we bring to life's great banquet?[7]

Changing: praising and communing with all creation

What scientists, original peoples and mystical traditions have been trying to tell us in a language we have not understood very well, is that all creation comes from one single source of life, wisdom and compassion (love that shares). God is giving himself to us in the grass that invites us to lie on it, in the dew that refreshes, in the wind, the cock crowing, crickets chirping, and in this miracle which is ourselves. The psalmist was right to cry:

> For so many marvels I thank you;
> a wonder am I and all your works are wonders (Ps. 139.14).

God shares himself with us in what we are, and in this gift of love that surrounds us, even in what we have thought of as our very own, our self-awareness. So it makes sense to share fully in life's festival, the praise of all creation.

We are not alone or isolated, we never have been. Because we are in communion with everything that is. We are the fruit of Love, Wisdom, the Life Force. From the spiritual desert of the sick human heart that is isolated, afraid, possessive, comes all the negative power (sin) which erodes and destroys itself, others and the very earth. Likewise, from the fruitful love of the open and wholesome heart springs praise, abundant life, gratitude and rejoicing in all that exists. Being converted to love means beginning to share more and more in the praise, adoration and communion of all creatures. Life and its continuation on the planet is only possible because it is a dynamic process of sharing, communing with one another.

This is what we mean by ecological balance. Only now are we beginning to understand that there is no waste in nature. Even death itself means life for the whole. From microbes to human beings, we are all designed for the subsistence of all. We could say that life truly consists of living in communion.[8]

This is the spiritual meaning of the incarnation of the Logos. In creation God breathes life into us. In the incarnation the Logos itself shares in the same matter of which we are made. It envelops itself in the dust which we are, and thus, as it were, enters into that constant ecological communion which is the life of the earth and all its creatures. The incarnation unites heaven and earth (Phil. 2): the incomprehensible mystery of life and the actual life of human beings on earth, what we see and what we cannot see.

This was Jesus' whole life: sharing and sharing himself. Because he shared bread, the fruit of men and women's labour, the fruit of the earth, he could call himself the bread of life. Even in the face of death he maintained this attitude of self-giving and gave himself up to it without resisting. Even after he rose from the dead, the disciples on the road to Emmaus sat down to table with him as a stranger, and recognized him when he broke bread and shared it.

The words of the cultural historian William Irving Thompson are illuminating here: 'When Jesus takes the bread and wine and says: "Take this in memory of me, because it is my body and blood", he is not the masochistic psychopath imagined by Freud, but a poet with an ecological vision of life, which uses myth and symbol to express that all life eats and is eaten by one another. The Upanishads express this idea in a different poetic language when they say: "The earth is food; the air lives on the earth; the earth is air; they are food for one another."'

Therefore St Paul's theological idea of Christ's mystical body is a vision of a planetary being, a single cell in which all of us are individual particles. If sharing food is the mainspring and source of our original humanity, then we really fulfil this humanity when we eat together.[9]

Wasn't this perhaps what Jesus was teaching us in the multiplication of the loaves and fishes? The first Christians understood that communion fundamentally meant sharing, being life for one another. This is not a matter of theology but of spirit. And it is a question of sharing not just the earth and its products but all the resources that a few people concentrate in their own hands, claiming they own them: information, knowledge, culture, points of view, methods, struggles, disasters and victories, worries and dreams: life.

Jesus is life sharing itself, resurrection and salvation. When people learn to share and share themselves, we will know what 'ecology of awareness' means. The life of our mother the earth depends on this change. And 'what happens to the earth will happen to earth's children'.

Notes

1. R. H. Stram and U. Oswald, *Por esto somos pobres*, Cuernavaca, Mexico 1990, 21, 23, 27, 35, 37.
2. W. I. Thompson, GAIA. *Implicaciones de la nueva biología*, Barcelona, 117–21.
3. Id., *Biología. La naturaleza vuelve a la vida integral nueva concienza*, Barcelona, 123f.
4. F. Capra, *The Tao of Physics*, London and New York 1983, 160, 167.
5. D. Zohar, *La conciencia cuántica*, Barcelona, 86–8.
6. Quoted in P. Russell, *La tierra inteligente*, Madrid 1992, 21.
7. A. Vittachi (ed.), *Simposium sobre la tierra*, Barcelona, 116.
8. Thompson, *La naturaleza vuelve a la vida* (n. 3), 103.
9. Ibid.

Liberation Theology and Ecology: Alternative, Confrontation or Complementarity?

Leonardo Boff

Liberation theology and ecological discourse have something in common: they stem from two wounds that are bleeding. The first, the wound of poverty and wretchedness, tears the social fabric of millions and millions of poor people the world over. The second, systematic aggression against the earth, destroys the equilibrium of the planet, threatened by the depredations made by a type of development undertaken by contemporary societies, now spread throughout the world. Both lines of reflection and action stem from a cry: the cry of the poor for life, liberty and beauty (see Exod. 3.7) in the case of liberation theology; the cry of the earth growing under oppression (see Rom. 8.22–3) in that of ecology. Both seek liberation: one of the poor by themselves, as organized historical agents, conscientized and linked to other allies who take up their cause and their struggle; the other of the earth through a new alliance between it and human beings, in a brotherly/sisterly relationship and with a type of sustainable development that will respect the different ecosystems and guarantee future generations a good quality of life.

It is time to try and bring the two disciplines together, to see to what extent they differ from or even confront one another, and how, basically, they complement one another. I begin with ecological discourse, since it represents a truly all-embracing viewpoint.[1]

I The ecological age

Ecology was originally seen as a sub-subject of biology, one which studies the inter-retro-relationships of living bodies one with another and with their environment. This is how its first formulator, Ernst Heckel, saw it in 1866. But then the fan of its understanding opened out into the three well-known divisions:[2] *environmental*, which deals with the environment and the relationships various societies have had with it, sometimes benevolent, sometimes aggressive, sometimes incorporating human beings in the environment, sometimes not; *social*, primarily concerned with social relationships as pertaining to ecological relationships, since we as personal and social beings are part of the natural whole and our relationship with nature moves through the social relationship of exploration, collaboration or respect and veneration in such a way that social justice (the right relationship among persons, functions and institutions) implies a certain operation of ecological justice (a right relationship with nature, equitable access to its resources, guarantee of quality of life); finally, *mental*, which starts from the realization that nature is not external to human beings, but internal, in our minds, in the shape of psychic energies, symbols, archetypes and models of behaviour that embody certain attitudes of aggression towards or respect for and acceptance of nature.

In its early stages, ecology was still a regional discourse, since it was concerned with the preservation of certain threatened species (the whales of the oceans, the giant panda of China, the golden myco-lion of the tropical forests of Latin America), or with the creation of nature reserves that would ensure favourable conditions for the various ecosystems. Or, in a word, it was concerned with the 'green' of the planet – with forests, principally the tropical ones that contain the greatest biodiversity on earth. But with the growth of consciousness of the undesirable effects of the processes of industrial development, ecology became a world-wide discourse. It is not only species and ecosystems that are threatened. The earth as a whole is sick and needs treatment and healing. The alarm was raised in 1972, with the Club of Rome's famous document, *The Limits of Growth*. The mechanism of death seems all-devouring: since 1990 ten species of living bodies have been disappearing every day. By the turn of the century, they will be disappearing at the rate of one every hour, and by then we shall have lost twenty per cent of all life forms on the planet.[3]

Ecology became the basis for a vigorous social critique.[4] Underlying the type of society dominant today is an arrogant anthropocentrism. We human beings see ourselves as being above other beings and lords over their life and death. In the past three centuries, thanks to scientific and technological advances, we have awarded ourselves the instruments with which to dominate the world and systematically plunder its riches, reduced in our minds to 'natural resources', with no respect for their relative autonomy.

The natural sciences developed in particular since the 1950s with the deciphering of the genetic code and the knowledge gained from various space projects present us with a new cosmology; that is, we have a coherent view of the universe, a different outlook on the earth and the way humankind functions in the evolutionary process.[5]

In the first place, we have gained an entirely new vantage point: for the first time in history, we, in the persons of the astronauts, have been able to see the earth from outside itself.[6] 'From the moon,' one of them, John Jung, said, 'the earth fits in the palm of my hand; there are no whites and blacks in it, no Marxists and democrats. It is our common home, our cosmic homeland. We must learn to love this wonderful blue-white planet, because it is threatened.'

In the second place, embarking on a spaceship, as Isaac Asimov recognized in 1982, on the twenty-fifth anniversary of the launching of the first Sputnik, makes it obvious that earth and humanity form a single entity.[7] This is perhaps the most basic intuition of the ecological approach: the discovery of the earth as a super-organism, given a name – Gaia.[8] Rocks, waters, atmosphere, life and consciousness are not juxtaposed, separated one from another, but have always been inter-related, in a total inclusion and reciprocity, making up one unique organic whole.

In the third place, we humans are not so much *on* the earth as *of* the earth. We are the most complex and singular expression known, so far, of the earth and the cosmos. Men and women are earth that thinks, hopes, loves and has entered into the no longer instinctive but conscious phase of decision-making.[9] The noosphere represents an emergence from the biosphere, which in its turn represents an emergence from the atmosphere, the hydrosphere and the geosphere. Everything is related to everything at all points and at every moment. A radical interdependence operates among living and apparently non-living systems. This is the foundation of both cosmic community and planetary community. We human beings need to rediscover our place in this global community, together with other species, not outside or above them. All anthropocentrism is out of place here. This does not mean renouncing our singularity as human beings, as those beings in nature through whom nature itself curves through space, irrupts into reflective consciousness, becomes capable of co-piloting the evolutionary process and shows itself as an ethical being which takes on the responsibility for the good destiny of the whole planet. As the great US ecologist Thomas Berry has shrewdly remarked: 'The final risk the earth dares to take is this: entrusting its destiny to human decision, granting the human community the power of decision over its basic life systems.'[10] In other words, it is the earth itself that, through one of its expressions – the human species – takes on a conscious direction in this new phase of the process of evolution. Finally, all these perceptions give rise to a new understanding, a new vision of the universe and a redefinition of human beings in the cosmos and of our actions in relation to it. Such a fact faces us with a new paradigm.[11] A new age has been founded: the ecological age. After centuries of confrontation with nature and of isolation from the planetary community, we are finding our way back to our common home – great, good, fruitful Earth. We are seeking to establish a new alliance with her, one of mutual respect and brother/sisterhood.

II Hearing the cry of the oppressed

How does liberation theology relate to ecological concerns? We have to recognize at the outset that liberation was not born out of the schema of ecological concern sketched above. The major challenge it addressed itself to was not the earth as a threatened whole, but its exploited sons and daughters, condemned to die before their time, the poor and oppressed.[12] This does not mean that its basic insights had nothing to do with ecology; they related directly to it, since the poor and oppressed are members of nature and their situation objectively represents an ecological aggression. But all this was worked out within a stricter historical-social framework and in the context of a classic cosmology.

The main thrust of liberation theology, back in the 1960s, was ethical indignation (the true sacred anger of the prophets) in the face of the collective poverty and wretchedness of the masses, principally in the so-called Third World. This situation seemed – and still seems – unacceptable to any basic human sensitivity and *a fortiori* to the Christian conscience, which reads in the faces of the poor and marginalized the actualization of the passion of the crucified Christ, crying out and longing to rise again to life and liberty.

The option for the poor against their poverty and for their liberty constituted and still constitutes the central axis of liberation theology. Opting for the poor implies action: it means putting oneself in the situation of their poverty, taking on their cause, their struggle, and – in the limit case – their often tragic fate.

The poor have never been the chief focus to such an extent in any earlier Christian theologies. The particular intuition of liberation theology was to try to build a theology from the viewpoint of the victims, in order to denounce the mechanisms that made them victims, and to help, with the aid of the spiritual heritage of Christianity, to overcome these mechanisms through the collective gestation of a society with greater opportunities for life, justice and participation.

This is why the poor occupy the central epistemological place in liberation theology; that is, the poor are the place from where to try and define the concept of God, of Christ, of grace, of history, of the churches' mission, the meaning of economics, politics and the future of societies and human beings. From the standpoint of the poor, we can see how excluding present-day societies are, or how imperfect democracies are, not to mention the religions and churches caught up in the interests of the powerful.

From earliest times, Christians have taken care of the poor (see Gal. 2.10). But the poor have never before been given such theological prominence, nor been seen to such an extent as transforming political agents, as they are by liberation theology. It never understood the poor in a reductive or merely 'pauperist' sense. The poor are not viewed just as beings in need, but as having desires, unlimited communication skills, a hunger for beauty. Like all human beings – as the Cuban poet Roberto Retamar aptly said – the poor have two basic hungers, one for bread, which can be satisfied, and one for beauty, which is insatiable. This is why liberation can never be sectionalized into material, social or merely spiritual segments. It is authentic only when the totality of human needs is kept open. It was liberation theology's merit always to have affirmed this integral character of human needs, from its first beginnings, from a right interpretation of what liberation means, not from any doctrinal demands from the Vatican.

The authenticity of liberation does not, however, consist merely in keeping its integral character, but also and principally in its being effected by the victims themselves, the poor themselves. This is perhaps one of the particularities of liberation theology by comparison with other practices from tradition that have also been concerned with the poor. Common understanding of the poor is of those who have not – food, housing, clothing, work, culture. Those who have, it is said, should help them to free themselves from their poverty. This approach is loaded with goodwill and right intentions; it underlies all assistentialism and paternalism in history. But it is neither efficient nor sufficient. It does not free the poor, since it keeps them in a régime of dependency: what is worse, it fails to appreciate the liberating power of the poor. The poor are not simply those who have not: they also have – culture, capacity for work, for collaboration, for organization, for struggle. Only when the poor trust in their own potential and opt for their like are true conditions created for authentic liberation. The poor make themselves into the historical agents of their own liberation; they also become free, capable of self-determination for solidarity with those who are not their like.

This is why we should stress that it is not the churches that free the poor, nor a beneficent state, nor the classes that assist them. These can be allies of the poor, provided they do not take their protagonism and hegemony from them. We can speak of liberation only when the poor themselves emerge as the principal builders of their own road, even if others help them build it.

One of the permanent merits of liberation theology undoubtedly stems from the methodology it introduced into theological reflection.[13] It does not start from ready-made doctrines, nor from revealed truth, nor from Christian traditions. All these are present on the Christian horizon, as a backdrop to illuminting convictions and as the flooring to reflection. But liberation theology starts specifically from the anti-reality, from the cry of the oppressed, from open wounds that have been bleeding for a long time.

Its first step is to accept reality at its most dramatic and problematic. This is the stage of *seeing*, of feeling and bearing the effects of human suffering. It means an overall experience of com-passion, suffering-with, protest-a[c]tion, of mercy and of a will to liberating action. This supposes a direct contact with anti-reality, an experience of existential shock.

The second stage is that of analytical *judging*, in the double sense: that of critical understanding and that of shedding light on the basis of the contributions of faith itself. We need to decipher the causes that engender suffering, seek their cultural roots, in the interplay of relationships of economic, political and ideological power. Poverty is neither innocent nor natural: it is produced, and so the poor are exploited and impoverished. The data of revelation, of tradition, of faith, of Christian practice down the centuries, denounce this situation as sin, that is, as something that also has to do with God, as a denial in history of God's design, mediated through justice, tenderness to the poor, sharing and community.

The third stage is that of transformative *action*, which is the most important, since everything has to result in this. It is important that Christian faith should make its contribution to the transformation of relationships of injustice into those that provide more life and joy in life, in sharing and in a reasonable quality of life for all. Christian faith has no monopoly on the idea of transformation, but joins in with other forces also taking up the cause and struggle of the poor, making its contribution with its religious and symbolic particularity, its manner of organizing the faith of the poor and its presence in society.

Finally comes the stage of *celebration*. This is a decisive dimension for faith, since it brings out the most gratuitous and symbolic aspect of liberation. In celebration, the Christian community recognizes that the specific achievements of its commitment are more than social, community or political dimensions. They are all these, but they also signify the anticipatory signs of the goods of the Kingdom, the advent of divine redemption mediated through historical-social liberations, the moment when the utopia of integral liberation is anticipated under fragile signs, symbols and rites.

Through its liberating commitment, based on theological reflection, Christianity has shown that the idea of revolution/liberation/transformation is not the monopoly of secular left-wing traditions, but can be a summons made by the central message of Christianity, which proclaims someone who was a political prisoner, was tortured and nailed to a cross as a consequence of his way of life, and who was raised back to life to demonstrate the truth of this way of life and to bring about the utopian realizations of the dynamisms of life and liberty.

III The most threatened beings in creation: the poor

We now need to bring together these two types of discourse, that of ecology and that of liberation theology. In its analysis of the causes of the impoverishment afflicting the majority of the world's population, liberation theology came to appreciate the existence of a perverse logic. The same logic of the ruling system, based on profit and social manipulation, that leads to the exploitation of workers, also leads to the spoilation of entire nations and eventually to the depredation of nature itself. We can no longer simply make technological corrections and redefinitions – though we still have to do so – in the style of reforms within this same logic; we need to move beyond this logic and way of seeing ourselves, which we have enjoyed for at least the last three hundred years. We can no longer go on treating nature, as present-day societies do, as a sort of supermarket or self-service cafeteria. Nature is our common heritage, which is being impiously plundered, but which we must conserve. We also need to guarantee the conditions for its later survival for our own generation and for future generations, since the entire universe has been working for fifteen thousand million years to bring us to the point we have now reached.

From being the Satan of earth, we have to educate ourselves to be its guardian angel, capable of saving the earth, our cosmic homeland and earthly mother.

The astronauts accustomed us to seeing the earth as a spaceship floating blue in interstellar space, bearing the common destiny of all beings. The fact is that on this earthship, a fifth of the population travels in the space reserved for passengers, and these consume eighty per cent of the provisions made for the journey. The other four-fifths travel in the cargo hold, suffering from cold, hunger and every other sort of deprivation. They are slowly becoming conscious of the injustice of this distribution of goods and services. They are planning to revolt: either we die passively of starvation, they tell one another, or we make changes that will benefit us all. The argument is not hard to understand: either we all save ourselves within a system of living together in solidarity and sharing with and in spaceship earth, or we explode it through our indignation and fling us all into the abyss. This understanding is growing all the time.

The latest arrangements of the world order ruled by capital under the regime of globalization and neo-liberalism have brought fantastic material progress. State-of-the-art technologies, those of the third scientific revolution, have enormously increased production. But the social effect is perverse: the exclusion of workers on a massive scale, and even of entire regions of the world, which are of little interest for the accumulation of capital in a cruelly indifferent mentality.[14]

Recent data suggest that total world profits are sacrificing the populations of Hiroshima and Nagasaki every day.[15] Progress is immense, but deeply inhuman. Its focus is not human beings and peoples with their needs and preferences, but merchandise and the market to which everything has to be subject.

In this context, the most threatened beings in creation are not the whales, but the poor, condemned to an early death. UN statistics indicate that fifteen million children die every year before finishing their fifth day of life, from hunger or the diseases associated with hunger. 150 millions are undernourished and 800 millions live permanently with hunger.[16]

It is from this human catastrophe that liberation theology starts when it meets the ecological question. In other words, it starts from social ecology, from the way human beings, the most complex beings in creation, relate to one another, and how they organize themselves in their relation to other beings in nature under régimes of great exploitation and cruel exclusion. What is most urgently sought is the minimum social justice required to ensure that life has its basic dignity. This presupposes more than social justice. It presupposes a new alliance between humankind and other beings, a new courtesy toward creation and the working-out of an ethic and mysticism of brother/sisterhood with the entire cosmic community. Democracy must become socio-cosmic: that is, the elements of nature such as mountains, plants, rivers, animals and the atmosphere must be the new citizens who share in the human banquet, while humans share in the cosmic banquet. Only then will there be ecological justice and peace on planet Earth.

Liberation theology should adopt the new cosmology of ecological discourse, the vision that sees the earth as a living superorganism linked to the entire universe. It should understand the human mission, exercised by men and women, as an expression of earth itself and a manifestation of the principle of intelligibility and loving care that exists in the universe; it should understand that human beings – the noosphere – represent the most advanced stage of the cosmic evolutionary process on its conscious level. They are co-pilots with the guiding principles of the universe that have controlled the whole process since the moment of the 'big bang' some fifteen thousand million years ago. Human beings were created for the universe and not vice versa, in order to bring about a higher and more complex stage of universal evolution.

Having adopted this basic stance, we need to define our starting point – an option for the poor that includes the most threatened beings in creation. The first of these is planet Earth itself, as an entity. Acceptance that the supreme value is the conservation of the planet and the maintenance of conditions in which the human species can flourish has not yet sufficiently entered general consciousness. This option shifts the axis of all questions; the basic question is not: What future is there for Christianity or Christ's church? Nor: What will be the fate of the West? It is rather: What future is there for planet Earth and for humankind as its expression? To what extent can Christianity with its spiritual heritage guarantee its collective future?

Then, we have to make an option for the poor of the world, for those immense majorities of the human species who are exploited and decimated by a small majority of the same species. The challenge is to make people see one another as members of a great earthly family together with other species and find their way back to the community of other living beings, the planetary and cosmic community.

Finally, we have to find a way of guaranteeing the sustainability, not of one type of development, but of the planet itself, in the short, medium and long term. This requires as a non-consumerist sort of cultural practice one that respects the rhythms of ecosystems, that produces an economy of sufficiency for all and delivers the common good not only to human beings but also to the other beings in creation.

IV Liberation theology and ecological discourse as a bridge between North and South

Two great problems will occupy human minds and hearts from now on: What is the fate and future of planet Earth if we prolong the logic of plunder to which our development and consumer model has accustomed us? What can the poor two-thirds of humankind hope for from the world?

There is the risk that the 'culture of the satisfied' will close in on its consumerist egoism and cynically ignore the devastation of the poor masses of the world. Similarly, there is the risk that the 'new barbarians' will not accept their death sentence and will launch themselves into a desperate struggle for survival, threatening and destroying everything in their path. Humankind could still be facing levels of violence and destruction never yet seen on the face of the earth, unless we – collectively – decide to change the course of civilization, shift its axis from the logic of means to exclusive profit to a logic of ends as a function of the common good of planet Earth, of human beings and of all beings, in the exercise of freedom and cooperation among all the nations.

Today these two questions, with different emphases, are common concerns of the North and South of the planet. And they make up the central content of liberation theology and of ecological reflection. These two thoughts allow for dialogue and convergence in diversity between the geographical poles of the world. They should be an indispensable mediation in safeguarding the whole of creation and in redeeming the dignity of the poor majorities of the world. So liberation theology and ecological discourse need one another and mutually complement one another.

Translated by Paul Burns

Notes

1. Cf. D. G. Hallman, *Ecotheology, Voices from South and North*, Geneva and Maryknoll, NY 1973.
2. Cf. F. Guatarri, *As três ecologias*, Campinas 1988.
3. See further data in L. Boff, *Ecologia, mundializaçao e espiritualidade*, São Paulo 1993 (ET in preparation).
4. Various, *L'écologie, ce matérialisme historique*, Paris 1992; Various, *Ecology, Economics, Ethics. The Broken Circle*, New Haven 1991.
5. Cf. M. Longair, *The Origins of our Universe*, Cambridge 1992; R. R. Freitas Mourão, 'Nature is an Heraclitean Fire: Reflections on Cosmology in an Ecological Age', *Studies in the Spirituality of the Jesuits* 25, New York 1991.
6. Cf. F. White, *The Overview Effect*, Boston 1987.
7. *New York Times*, 9 October 1982.
8. J. Lovelock, *The Ages of Gaia: the Biography of Our Living Earth*, New York 1988.
9. E. Jantsch, *The Self-Organizing Universe: Scientific and Human Implications of the Emerging Paradigm of Evolution*, New York 1980.
10. *O Sonho da terra* (The Dream of Earth), Petrópolis 1991, 35.
11. R. O. Muller. *O nascimento de uma civilização global*, São Paulo 1993.

12. Cf. H. Assmann, 'Teologia da solidaridade a da cidadania ou seja continuando a teologia da libertação', *Notas de ciências da religião* 2, 1994, 2–9.

13. See the already classic work by C. Boff, *Teologia e prática*, Petrópolis 1993.

14. Cf. F. J. Hinkelammert, 'La lógica de la expulsión del mercado capitalista mundial y el proyecto de liberación', *Pasos*, San José, Costa Rica 1992.

15. Cf. R. Garaudy, *Le débat du siècle*, Paris 1995, 14.

16. Cf. UNDP, *Human Development Report*, Oxford 1990.

Some Premises for an Eco-Social Theology of Liberation

José Ramos Regidor

Liberation theology came into being as critical reflection on the faith experienced by Christians in base communities in Latin America in solidarity with the great majority of the poor. From their perspective, it has made critical use of the socio-political and humane sciences in order to understand the mechanisms of the model of colonial development which is produced by this social injustice, by listening to questions from this reality and by discovering the role of believers in the struggle for liberation.

The ecological crisis arose in the northern hemisphere more than 150 years ago, as a consequence of the industrial revolution. Therefore ecological culture came into being and has developed in the context of a prosperous society. However, for about twenty years the ecological crisis has also become visible and has proved devastating in the countries in the southern hemisphere, in the context of a society of wretchedness and poverty. So the dominant ecological culture of the North cannot by itself understand the interweaving of the social crisis and the ecological crisis, the struggles for social justice and the struggles for respect for and liberation of nature, in other words for ecological justice. It is in this context and with these questions that social ecology, political ecology and so on are emerging in the South.

In this perspective it is possible to develop an eco-social theology of liberation. Making critical use of the new ecological culture of the South and some contributions from the ecological culture of the North, it has to present the perspective of the South, in solidarity with the social and environmental struggles of the South, while remaining open to questions from believers and seeking to discover the role of Christian faith in these historical processes.

In 1994, after a series of other articles, Leonardo Boff published a systematic study on the challenge of ecology to society and to Christianity. Beginning from the South, he analysed the emergence of a new culture and a new planetary awareness. Always in the context of the ethical and political perspective of the South, he spoke of the encounter between the new global order (disorder) and Christianity, with an emphasis on a utopian horizon.

In analysing the current historical processes, we can begin with some of the main conflicts: between North and South, between human beings and nature, between men and women. Any of these problems can help towards developing a general and contemporary view of the world. However, in general, they are interpreted from the dominant perspective, which is that of the culture of the North: in relations between the North and the South the North is celebrated as being able to impose a model of development on the whole world, with a Eurocentric, colonial and dominant attitude. In the relationship between human beings and nature, human violations of the earth and its eco-systems (geocide, ecocide) are justified; in relations between women and men the masculine perspective prevails.

So, as a premise for the development of an eco-social theology of liberation it is necessary to reject the perspective of the dominant culture and choose the point of view of the victims, those who have been excluded, in order to join in discovering a different interpretation of the three main conflicts to which reference has already been made, with the aim of helping to shape a common destiny towards an alternative civilization.

I From the perspective of the South

In the 500 years of conquest-colonization, the West tried to impose everywhere the model of civilization which was in process of development, namely modernity. This is a reality which has both positive and negative aspects, like every culture and every type of society. In the many attempts to arrive at a deeper understanding of it, we need to remember that modern European and Western civilization has always had two faces, above all in its relations with the South. On the one hand, it has a rational, critical and emancipatory dimension and presents itself to other peoples as a culture of democracy and freedom, of rights and human dignity, of the lay state, of the development of science and technology, of the spread of a degree of prosperity, and so on. On the other, it also has a dimension of universalizing and exclusive rationality which is based on the myth of its superiority and its position as the centre of the whole world, as a kind of absoluteness to which it has sacrificed, sometimes without being aware of the fact, the cultures, peoples, men and women of the worlds that it has colonized. This dimension is the root of the different types of exploitation, exclusivism and racism, of genocide (the destruction of peoples), biocide (the destruction of life), ecocide (the destruction of eco-systems) and geocide (the destruction of the earth).

So it can be said that one dimension of modernity expresses the historical experience of the colonial system as experienced and practised by the conquerors and the colonizers from the dominant sectors of the North and West. Another dimension of this modern civilization expresses the experience of 500 years of colonialism from the perspective of the South, of the victims, of those who have been excluded. However, the dominant culture and historiography of the northern hemisphere try to ignore and hide these perspectives of the conquered, despite the progress of anthropology and ethno-history in a critical evaluation of their sources. In any case, it is ethically and historiographically incorrect to reject the awareness of either the colonizers or the victims, who are all part of human history.

The colonial system in the broad sense is based on the creation in those lands of a social, political and cultural organization capable of ensuring the flow of resources and capital towards the North. This essential objective of European colonial expansion (then that of the West and the North) was always achieved with the support of the colonizing countries for the dominant elite in the countries of the South. These countries were given the task of controlling the great impoverished masses (amounting to around 80%) who were largely excluded from the system, also with the intervention of the army and the police. The North assured these dominant sectors a level of prosperity similar to that of the countries at the centre, not least because the sectors had the function of being the favoured markets of the North.

One of the main results of this colonization and modernization has been the accumulation of wealth in the North and among the dominant elites of the South. All in all, they form about a fifth of humankind and consume four-fifths of available resources. There is therefore an unjust imbalance, unjust because it is intolerable for those who endure it and because it sets the privileges of the few above the survival of the many.

This accumulation of wealth has been produced by the model of development which has taken shape in these 500 years, one of the main dimensions of which is that of exclusion from work, from society, from culture and so on. In the form of society which exists at present more than a billion people are really poor because they are excluded from the cycle of production and consumption which is controlled by the power of the North. Furthermore, while the computerized, automated and universalized productive apparatus produces more things better, at the same time it necessarily excludes the workers and the southern hemisphere. So this mechanism of development has produced and still produces an unfair accumulation of wealth at the centres of capital, in the powerful sectors of the North.

For 500 years the peoples of the South were unable to be protagonists in history. Today, the flood of immigrants means that they are breaking into history with two objectives: the claim to be subjects and protagonists, and the organization, together with the North, of a fair and equal distribution of the wealth accumulated in the North. Furthermore, as became evident in the indigenous insurrection of 1 January 1994 in the state of Chiapas, Mexico, in many situations the Indios have become active subjects. They are emerging in history as protagonists and no longer want gifts or concessions imposed from on high, but to particpate in and contribute to a common destiny. A continental campaign was launched, entitled '500 Years of Indian, Black and Popular Resistance', centred on the debate and the demonstrations which marked the quincentenary. To prevent these peoples dropping out of view again, the Guatemalan Indio Rigoberto Menchu, winner of the Nobel Peace Prize in 1992, has tried to organize an 'International Decade of the Rights of Indigenous Peoples', from 10 December 1994 to 31 December 2004, which he launched with the support of the United Nations Organization.

The foreign debt of the countries of the South takes on another significance in this perspective. First of all it is a fact of economic finance, which the North maintains as an instrument of political control. From the perspective of the South, i.e. in the light of the 500 years of conquest and colonization, it has quickly been understood that over and above the economic and financial debt there is also an ethical historical and ecological debt which has many dimensions (socio-economic, political, military, ecological, cultural and religious). From this perspective the peoples of the South have moved from being debtors of the North to being its creditors, by virtue of the losses that the latter has caused and the resources and capital which it has sequestrated. To make possible some compensation of the peoples and of nature, it has also been proposed that the debts contracted by the North over against the South should to some degree be quantified. Above all there has been the experience that the rigidity of the economic theory of the North is hindering the removal of these unjust mechanisms. But at all events, this ethical-historical and ecological debt is an indelible sign of the unjust situation and thus of the ethical responsibility of the rich of the North and the South.

The topic of the distribution of wealth has also been taken up by Leonardo Boff. He begins from a living wage guaranteed to all. And he adds that according to some theorists, a survival wage should be granted to the two-thirds of the world population who are poor and marginalized: 'Indeed, this initiative is quite plainly an ethical duty required of the rich countries which at one time were colonizers. By exploiting the colonies, they gained the first accumulation of resources for the leap towards modernity and industrialization. As a just imperative and not a matter of social charity, the rich countries should grant a survival wage to those who today are poor and yesterday were the colonized.' That is necessary in order to maintain the unity of humankind, without a war of survival among the two-thirds of the South who are poor and do not peacefully accept the death penalty which hangs over them. This would call for a global change in the world economy; it would need to be redefined so as to be put at the service of society and nature. Boff speaks of a new economic paradigm, the 'multidimensional economy', orientated on a distributive justice which consists in giving to each according to their needs, their talents and their work and thus relativizing the economy of trade.

So to take the perspective of the South means to fight Eurocentrism. That does not mean to deny Europe, but to criticize and eradicate from the dominant collective imagination that view of the world in which the North appears superior and the centre of all, a view which is shared by the mass media and internalized in the consciousness even of the colonized countries.

The organization of the economy in the indigenous communities of Latin America quickly brings to light the presence of a culture which is different from that in the West. For these communities the economic problems are important, but lie within the ethical dimensions of their cultural traditions, above all in their mode of relating to the earth. This is a different way of living and producing from the economies of the societies of the North, but it could be a stimulus to the shared quest for a society in which the economy is not the dominant paradigm. It should be possible to create social, political and cultural conditions which allow those who have been excluded to be put on a par with the rest of humankind and not left on the periphery. Once this option has been taken, it is a matter of creating together an economic project of world solidarity which favours the life and dignity of all and not the profits of some. This is a radically different project, which could be a first step towards an alternative culture to that of colonialism.

That will be possible if these economic projects are inspired by recognition of and respect for the subjectivity and otherness of individuals and nature. This is a necessary premise for a democratic culture which is eco-social and multi-ethnic. There should in fact be an awareness that all cultures and all religions are relative historical realities, not absolutes; are partial and not total. At all events, only the awareness of our limits can arouse in all of us the capacity which is needed to achieve a mutual recognition of the worth of others, of what we have in common and of the relative differences, in order to lead to the enrichment of all, including our relations with nature. Otherwise Western civilization will continue to be a religion, a culture and an economy of domination, of exploitation and of death, as happened in the 500 years of colonialism down to the present. And the non-Western cultures and religions, too, if they remain absolutist, considering themselves to be the sole possessors of the whole truth, will continue to be imprisoned in the violences of their fundamentalisms. All this means that fighting Eurocentrism also means fighting against any type of ethnocentrism, and therefore fighting against those fundamentalisms which have their roots in a belief that they are in possession of the absolute, that they are superior and at the centre of history.

These reflections lead to a brief clarification of the terms South and North. The South is no longer solely a geographical reality but also a condition of life, which is also present in different forms in the rich countries of the North. 'South' denotes that condition of profound poverty and suffering which goes with the life of those who have been excluded. This is a condition which is often experienced with great simplicity and deep sensitivity, full of wisdom and human warmth, of solidarity and generosity, characteristic of the culture of poor peoples, but also with negative aspects which are produced by their wretched situation. The North, too, is no longer just a geographical reality but a condition of life, marked out by the industrial and technological system, by the ideology of progress and economic growth, by money, by consumerism, by efficiency, by power, by racist closedness to others, and so on. As has been said, there are also sectors of the North in the countries of the South, more or less bound to the ruling powers, with the task of making the interests of the colonizers present in the South. Furthermore, in the countries of both North and South there are sectors which contest the political governments and are in solidarity with the poor sectors of the South and the North. However, the geographical dimension retains a degree of validity: being poor or in solidarity with the poor in the geographical South is not the same as it is in the geographical North; nor is having a place in the dominant sectors of the geographical South the same as it is in the geographical North.

II From the perspective of nature

The ravaging of nature by human beings began in the neolithic age with the spread of agriculture, the first settlements and the first cities. The last 500 years, with European colonial expansion and mercantile culture, have seen the first forms of systematic exploitation of the earth. Finally, the mass destruction of nature which has produced the present ecological crisis began around 150 years ago, with the establishment of science and the technological revolution.

This exploitation of nature is justified by a dominant cultural tradition which was present in the patriarchal societies and which in the West was developed by Descartes, Bacon, Newton and so on. Its anthropocentric (patriarchal, capitalist and socialist) paradigm contrasted man with other beings by keeping him outside and above nature and putting him at the centre, as lord and master. In this perspective, nature is seen as an object and as an inexhaustible resource and therefore as susceptible to being dominated and exploited endlessly, ensuring an economy of growth.

However, there is another cultural tradition, present but not dominant in modern civilization, i.e. the ecological paradigm, which is emerging once again. From this perspective, at the global forum in the World Summit on the Earth concerned with the environment and development in Rio de Janeiro (June 1992), involving non-governmental organizations, it was said that there is a need to move from a type of ecology which is concerned with all beings, living and non-living, human and non-human, in their reciprocal interdependence. In fact a recent history of ecology (by J. P. Deléage) argues that today we need to speak of ecology as a 'science of nature and human beings' and not just as 'a science of nature'. Along the lines of this approach, different forms of social, political, human and global ecologies are coming into being, and a series of sciences are being rethought from an ecological perspective: eco-economy, eco-sociology, eco-politics, eco-psychology, eco-theology, and so on.

According to this conception, nature is not totally outside but within human beings. Human beings, as living bodies, are products of nature, of which they form a part. From this perspective it is necessary to distinguish between internal and external nature. Internal nature is our physical body with its biological life, its psyche, its experiences, its sexuality, and so on. The nature which is external to our bodies is nature as environment, i.e. the totality of other beings present on the earth on which they all live. These are interdependent, so that their relationships give rise to culture which is neither opposed to nor separate from nature but is in a situation of reciprocity. However, as thinking bodies, human beings are distinct from all other beings because they are the only ethical subjects capable of understanding the principle of nature and of themselves, capable of deciding by taking into account the pros and the cons, and also capable of adopting the point of view of others. Precisely by virtue of this unique characteristic of theirs, women and men are responsible towards nature and themselves. To some degree they can promote or destroy the planet. As Leonard Boff remarks: 'Humankind, above all with the advent of the industrial revolution, has proved to be an exterminating angel, a Satan of the earth. But it can become a guardian angel; it can help to save it, because it is its homeland and its mother country.'

Thus from the perspective of nature the countries of the North and also those of the South have a debt to nature, an ecological debt, for the damage produced above all by a technological economic situation which has organized its exploitation by using limitless violence. This mode of production and consumption has created mechanisms of pollution and degradation of the biosphere which threaten life on earth at a local and at a planetary level. These mechanisms include the greenhouse effect produced by gases emitted into the atmosphere and the consequent warming of the climate; the reduction of the ozone layer; desertification and the destruction of forests; the pollution of earth, water and air produced by fertilizers, pesticides and detergents; the storage of toxic waste and chemical and nuclear contamination. In particular, it needs to be emphasized that the North is the main culprit here. First of all because as the centre of the industrial and technological system and of consumer society it produces 80% of world contamination. It is true that this debt to nature also exists in the more or less industrialized countries of the South. And it is also true that in the South there is a partial destruction of the forests brought about by those peoples who, on the basis of the model of development imposed by the North, are forced to use them without protecting them. But it is the industrial system of the North which builds dams and hydro-electric power stations, establishes industries using precious woods, mines and cattle-breeding centres and so on which is destroying the tropical forests at a rate 70% greater than the complex destruction produced by the countries of the South. Finally, the North exports toxic waste, chemical and nuclear, to the South, with serious ecological consequences for the peoples and the natural environment of the South.

To overcome the anthropocentric paradigm characteristic of this model of development, it is necessary to work out a new paradigm, from the perspective of nature – an ecological paradigm. This means above all the recognition of the otherness of nature, in other words, recognition of and respect for the intrinsic and autonomous value of every being on the planet, quite apart from their relations with human beings. It needs to be possible for human beings to use non-human beings without destroying the ecosystems, in solidarity with them, recognizing their limits, the laws of their equilibrium, their positive language and potentiality. In particular, it is necessary to create a culture of limits and quality of life, capable of overcoming the ideology of unlimited quantitative growth and creating a new type of economy which does not give priority to profit and unlimited consumption, because that would increase the deterioration and finally bring about the death of all beings, human and non-human, today and for future generations.

To cope with these and other problems, some years ago an ecological economics developed, understood as a new discipline centring on the interface between economic research and ecological research. This new science has been interpreted in different ways. It frequently refers to the debate on sustainable growth. Others, with its theoretical support, assert that to the Marxist conflict between capital and labour should be added the second conflict between capital and nature. These approaches are present in part in the research of the peoples of the South. However, the majority does not reject unlimited quantitative growth and remains within the coordinates of the capitalist and socialist industrial system and their economistic paradigm. Some argue that the new technologies will be able to resolve the problems of the environment, with the risk of eco-business and the commercialization of nature. Global ecology analyses above all the problems of environmental policies, at the local and planetary level. It emphasizes the difficulties and warns against the tendency of the political summits of the world to manipulate ecological problems so as to create forms of ecocracy and technocracy in place of democracy.

Certainly the ecological debate is very rich and articulate. In the most critical sectors, it seems necessary for there to be a new organization of the economy which is not economistic but bound to ecology and to nature, to the service of the communities and of nature. There is a need to begin from the recognition of the otherness and the limits of nature and therefore from the rejection of unlimited economic growth. This new beginning must be capable of creating a kind of limitation of products, of consumption and styles of life which can ensure a fair distribution of the wealth and qualities of life for all in both North and South. There is need for a radical change of the present mode of development and civilization.

The topic of the equal distribution of benefits, 'albeit limited', in the North and in the South is connected with what is said about the debt of the North and the South to nature. This is a debt which differs substantially in the two cases, because the responsibility of the North is much greater than that of the South. The social question and the ecological question intersect. Those sectors of the North which have become aware of the problems have pledged themselves to change their forms of production and their life-styles, to limit the use of energy and to boycott the consumption of those goods which lead to the impoverishment of the peoples, including those living a long way away, and the plundering of nature. May we ask for a similar option and practice from those sectors of the South which have similarly become aware of the problems? For the poorest, the limitation of production and consumption is a question of survival. It is true that the desertification of the fields and forests and the growing pollution of the major cities of the South is increasing an awareness of the ecological crisis. But the North must not impose environmental policies which have been developed in its own interests. Furthermore, the North must not impose the categories of the dominant ecological culture, which came into being in a situation of prosperity and abundance. In the South the ecological crisis goes along with the social crisis. It is therefore a task for the South to create a new culture and a new ecological policy, developed in context, as is happening with the rise of social ecology, political ecology, and so on. We can ask for a dialogue on equal terms to struggle together to change this form of civilization. In it there must be a presentation of the perspectives of the South, of nature, and of women.

III From the perspective of women

Since the last century, the feminist movements have laid claim to the perspective of women, asserting their capacity to construct a global vision of the world that can express the experience of the half of humankind which so far has been unable to manifest it.

In a first phase they struggled for equality and parity with males, for status and involvement in social and political life. In the last twenty years the recognition has been established that they are different in sex and gender. This first of all means the claim and the recognition of the right of women to their self-determination, to control of their own bodies, their sexuality and their fertility.

However, in the situation of the poor women of the South this free affirmation of their sexuality has been hindered and denied by the patriarchal, colonialist and capitalist structures which are still dominant at an economic, political and cultural level. As the women's movements of the South repeated in Cairo, for the poor women of the South to be able to make a free choice in connection with their sexuality and their fertility there is a need to establish the social, economic and political conditions for this freedom. And in the case of the women's movement as well, that calls for commitment to the construction of a new model of development and society.

Some women's movements in the South see the beginning of a third phase of the women's movement in that the recognition and the exercise of differences of sex and gender includes the responsibility to be concerned not only for individuals but also for the world in which they live, by presenting and struggling for a new type of society. Here, then, it is a question of living on the interface between sexual difference and responsibility for the world. According to the Brazilian feminist Rosiska D'Arcy de Oliveira, feminism has become a 'historical context for thinking and acting in the world', from the perspective and with the content of women, with the proposal of a 'civilizing break' capable of taking up the achievements of parity. Of course this will be in dialogue with men and with other movements and social groups aiming for an alternative, with the recognition of differences without hierarchy.

The women's movement, too, was born in the northern hemisphere, in the context of a developed and prosperous society. However, among the movements of the women of the South a new culture of women is in process of creation, with reference to the situation of the poor women of the South. Their eulogy of the difference represents a stimulating contribution to the debate on women, ecology and development and also represents a challenge to the women and the men of the North and the South. It is a matter of deepening the criticism of the male paradigm, the anti-ecological and Eurocentric paradigm, in order to establish in new forms the interface between social questions (the impoverishment and negation of peoples), environmental questions (the otherness of nature) and women's questions (subjectivity and difference). However, it should be noted that the male paradigm and the feminine paradigm are always relative and linked realities, with positive and negative aspects. Sometimes the feminine principle and the male principle are said to be realities present in both women and men. Vandana Shiva, a scientist who is a member of the women's movement in India, has developed this topic in connection with the ecological question and with the changes in the present model of development needed.

IV A critical break

The previous pages have offered a criticism of the colonial system of Western civilization, its paradigms of economism and Eurocentrism, of the anthropocentric plundering of nature and of masculinity. These systems and mechanisms produce injustices and risk making present-day civilization quite rigid, leading to a type of regime which is a fortified citadel, or to a global apartheid. There is the risk of a socio-political organization to defend and serve the prosperity of the North and the elites of government in the South, supported by colonialist powers and using racist laws, rigid economic policies and sophisticated military instruments to control those who have been excluded, in the South and in the North. There is the risk of a barbaric capitalist system, with an economy of exclusion and death, a system which can also lead humankind to destroy itself as a consequence of a possible ecological disaster and the use of chemical or nuclear weapons. If this model becomes dominant, the end of modernity could coincide with the destruction of human life.

It is true that today the popular movements for liberation which appeared in the 1960s and 1970s are no longer present. But it is also true that for about ten years, in civil society, a breath of hope has been felt at the grass roots and has been fanned by a great variety of new groups and associations, more at the level of micro-projects, committed to the struggle for justice, for human rights, for nature, for the recognition of the difference of women and their new subjectivity and otherness, for peace, for eco-social democracy, for health and food, for an alternative mode of communication, and so on. In the South and in the North, in the East and in the West, these groups and movements coming from the grass roots are seeking together to construct a new type of civilization which begins from the perspective of the South, of nature, of women.

Their demands call for the abandonment of the economy as the main paradigm and the emergence of a new paradigm. In other words, they seek a new point of reference which is not only economic but also cultural, ethical and political. Perhaps it will be necessary to discover the convergence of different paradigms and to become committed to the creation of an ecologico-economico-feminist and cultural paradigm, without any claim that this is the one and only paradigm, in an awareness that any historical process is always relative, limited, lay.

This change of civilization is understood as a 'civilizing break'. It is a 'break' because those bringing it about are not the powerful of the world, but the least: those who are excluded, women, nature, those who formerly were on the underside of history and could not be protagonists. Precisely for this reason, here we have a radical alternative to Western civilization, whose representatives are no longer able to find adequate responses to the problems which threaten the whole of humankind. It is 'civilizing' because it accepts dialogue and a critical revision of the achievements of modernity, seen from the South, from nature, from women. This calls for great strength in rethinking and renewal, and if necessary in abandoning the categories and the theoretical developments which are at the centre of the civilization that is moving into decline.

V Some suggestions for theological reflection

The religions, the churches and the theologies are part of the historical process which I have described. As far as Christianity is concerned, official Christianity has been an accomplice and an agent of colonialism and favours a culture of domination. The Catholic Church today is experiencing a marked centralism and Romanization, often proving incapable of incorporating the values of freedom, pluralism and democracy. It is dominated by the perspective of the powerful, by Eurocentricity and the identification of the gospel of Jesus Christ with its European version.

However, the Christianity of the base communities, to some degree in contradiction to official Christianity, is trying to create from the base a new type of Christianity, presenting it from the perspective of the South, of nature and of women. In solidarity with these new historical subjects, Christians in the base communities are rejecting the economistic and Eurocentric paradigms, the paradigms of violent dominion over nature and of masculinity. Positively, they are seeking to play a creative role, supporting the relativization of every idol and every master, in the lay processes of the creation of a new ecological-economic-feminine and cultural paradigm and seeking a new type of civilization based on the creation of a social, ecological, feminine and multi-cultural democracy.

In this perspective, Christians of the base communities are becoming aware that our everyday style of life, our way of living and thinking, of production, consumerism and waste is based on unjust relations which lead to the impoverishment of those who have been excluded, the degradation of the biosphere and discrimination against women. This awareness of the planetary dimension of our everyday life, where it is part of the mechanisms of the globalization of the economy and of politics, is a challenge to our responsibility and our Christian faith.

Simply from the perspective of faith, this reality can be interpreted as social, ecological and eco-social sin. For liberation theology, in addition to the individual sin committed by individuals, there is also a social sin which is the responsibility of all (citizens and society), and which is committed through the structures of the present model of development that produce social injustice against poor people. In parallel, it is possible to talk of ecological and eco-social sin, to the degree that we are all responsible for mechanisms of this type of development which threaten the possibility and the quality of life on earth. The option for the poor can be broadened so that it becomes the option for posterity, for the future generations who will suffer the inhuman consequences of the degradation of the biosphere, within a process which is beginning to set in motion perverse mechanisms bound up with our life-style that we can perhaps still block. The biblical commandment 'You shall not kill' (Exod. 20.13) refers primarily to individuals and peoples living at the present time, but it also refers to those coming after us who will be affected by mechanisms which are taking shape in the present and on which our everyday options have an effect.

These reflections have a place within research into a new eco-social ethic which has its main foundation in the recognition of others, in the subjectivity of the excluded, in the otherness of nature and the different status of women. This is an ethic capable of making an appeal to personal conversion, to a profound change in the predatory, colonialist and masculine, competitive and consumerist mentality which has created our life-styles and justified the values of everyday life. It is a call to changes in society, to the radical transformation of social and political structures and therefore of ways of living and producing which result in the impoverishment of the poor and the degradation of nature. It is a call to move from the logic of dominion and efficiency to the logic of free and effective love, from colonial racism to an acceptance of others, from the logic of unlimited quantitative growth to the culture of a respect for limits as a necessary condition of the quest for new qualities of life. Finally, it is a call to commitment to social and ecological justice so that the life of the poor and the life of nature is promoted and respected.

In this perspective an attempt is being made to develop an eco-social theology of liberation, understood as a critical reflection on faith as experienced by believers and their communities in the struggles for social justice, for ecological justice and for justice for women. Therefore the option for the poor must be experienced within the more general option for life, for the integrity of creation, especially where life is most threatened, i.e. among those excluded from the system, in nature, among the poor women of the South. This will be a theology which, as has been said, will broaden the option for the poor to an option for those coming after us, who can become the future victims of our everyday behaviour. Indeed the liberation which Christians in Latin America seek is at the same time liberation from the de-humanization and oppression of the poor (social liberation: the indigenous, the Blacks, sectors of the population, future generations), liberation from the corruption of nature (ecological liberation, a move from plundering to the recognition of living and non-living beings) and liberation from discrimination against women, a move towards the recognition of self-determination and the difference in sex and gender, above all for the poor women of the South. Only by struggling together for these objectives will we be able to build a society with fewer poor and more life.

The eco-social theology of liberation has the task of rethinking and reformulating the traditional theology of creation, presenting it from the perspective of the South, of nature and of poor women. There needs to be an awareness that this theology has necessarily made use of a culture marked by economism, masculinity and Eurocentricity. Here I shall suggest just two themes.

1. The category of creation calls for an overcoming of the dualism between human beings and nature. Therefore, as I have remarked, nature is not only external to human beings but stands within them, because we are all part of nature, all part of creation: we are all creatures. Every creature has its own intrinsic value, its role, and all exist in a reciprocal interdependence. The integrity and dignity proper to every being derives not from any reference to humankind but from a reference to the love of God for every creature. From this perspective, the 'debt to nature' which has been spoken of above is a debt to creation, a debt to God, an offence against God's love which is present in every creature. This approach sets out to overcome anthropocentricity and to recognize and respect the otherness of nature, of every creature, living and non-living, human and non-human.

2. Creation is not a static reality, as an event which happened once and for all. Rather, it is a historical process in which the creator has entrusted to human beings the responsibility of administrating creation. They are called to look after, to promote and complete the whole of creation. Human beings, especially in the last 500 years, have become 'a true Satan of the earth', which has been ravaged and corrupted in the service of the interests of the North. They have introduced socio-ecological sin, injustice and social and environmental violence in addition to personal violence, understood as the violation of the subjectivity and otherness of individuals. But as St Paul remarks (Rom. 8.18–25), the promise of the resurrection extends to the whole of creation, which today is already sharing in the liberation contained in the promises. So human beings can intervene in the historical process of creation to take care of it, reintegrate it and liberate it from the corruption which has been introduced by sin. Therefore the struggles for personal justice, for social justice and for environmental justice are inseparable.

Today the churches and the religions are confronted with the challenge of the emergence of new cultures and religions and therefore the quest for a society based on dialogue between races, cultures and religions. At the time of the conquest of America, Bartolomé de Las Casas and other missionaries and theologians tried to adopt the perspective of the Indios, thus recognizing the existence of a different perspective from that of Europe. In recent years liberation theology has been much occupied with the question of the interaction of evangelization, culture and religion. This problem was at the centre of discussions at the Fourth Conference of the Latin American Bishops which took place at Santo Domingo between 12 and 28 October 1992. Leonardo Boff and other theologians close to the base communities asserted that so far there had not been a true proclamation of the gospel, but the authoritarian imposition of the European gospel. There was therefore a need to distinguish between the gospel message of Jesus and its expression in a particular culture with which it was not to be identified. So far, evangelization had been identified with Westernization. Therefore it was necessary to note that European Christianity is not the only possible kind of Christianity and to recognize the right of the Indios and the Afro-Americans to their own religion, which was different from Christianity. It is in this perspective that the more or less developed reflections of indigenous theology, of black or Afro-American theology, and so on are now emerging. Furthermore, the ecological crisis and the situation of the poor women of the South are a challenge to the Christian base communities in Latin America: here various theologies of women's liberation, eco-theologies and so on are emerging.

These basic realities, for all their limitations, are a sign of hope. They offer hope because they are at the interface between the criticism of present-day civilization and the creation of a new Christianity among base communities, within the quest for an alternative civilization, from the perspective of the South, of nature and of women.

Translated by John Bowden

III · New Horizons

Towards a New Paradigm of Production: From an Economy of Unlimited Growth Towards One of Human Sufficiency

Experiments in Communitarian Production and Ecological Equilibrium

Bastiaan Wielenga

The problem posed in the topic assigned to me is of urgent global relevance. But it has to be discussed on the level of local experiments. This article takes its main examples from India, where the need for a new paradigm arises in the context of survival struggles of marginalized masses. The need for a new paradigm is equally urgent in the centres of the capitalist world. But experiments and political projects may differ.[1]

I The old paradigm is not viable

The old paradigm of unlimited growth has fundamental limitations which make the present economic system unviable.

1. Global ecological crisis

Environmental problems have occurred since the dawn of human civilization. But today's global economy with its never-ending drive for capital accumulation as the motor of unlimited growth has for the first time created ecological problems on a global scale which are threatening to engulf the planet and humanity as a whole, and the poor first of all. The rapid depletion of non-renewable resources, the greenhouse effect of the burning of fossil fuels, the poisoning of food-chains, the destruction of life-sustaining bio-diversity and a whole range of connected problems indicate that the present pattern of industrial and agricultural production, global division of labour, intensified world-wide trade and feverish consumerism cannot be sustained for long, and even now is sustainable only for a minority.

2. Mass poverty

Economic growth through endless technological innovation has increased productivity, but has not brought an end to poverty. Concentration of capital, investment in large-scale high-tech projects, automatization of industrial production, mechanization of agriculture and rationalization of the service sector result in marginalization and redundancy of a growing number of people. At an earlier stage of the 'constant revolutionizing of production' (Marx), uprooted subsistence producers could find a – miserable – place in the ranks of the industrial proletariat. And others were allowed to emigrate. These options are closed for those who today are driven from their lands and out of their artisanal occupations. Whatever globalization may mean, they are left out to struggle for survival on the margins of society, an outside class of redundant people who do not count as far as capital is concerned.

3. Socio-cultural disintegration

Marx spoke of the 'uninterrupted disturbance of all social conditions' as a result of the ongoing revolutionizing of production in the pursuit of accumulation. 'All that is solid melts into air, all that is holy is profaned.'[2]

Uprooted from local communities and traditional occupations, people face the loss of traditional status, of cultural meanings and social identity. Many may find some solace in the opiate of vulgar entertainment provided by global media around the clock. Others may find refuge in so-called fundamentalist movements whose upsurge reflects in a politically distorted way the need for cultural identity and something to fill the social and ideological vacuum. Modern consumerist capitalist society offers no solution.

II In search of human sufficiency

1. The precious tenacity of marginalized producers

Long ago, liberals and Marxists had declared subsistence producers, small peasants and traditional artisans to be obsolete. They would not survive the onslaught of capital. They would have no place in the future society. Many have been subordinated to modern production processes, many have perished. But others are still there.

The upsurge of indigenous peoples shows that even the most brutal, genocidal methods of liquidation have not succeeded completely. Still there are survivors who now confront the world threatened by eco-catastrophes with the relevance of their ways of life and patterns of production. Their need-orientated economy of 'enough' and their symbiotic ways of relating to nature are not just a sad reminder of what has been lost for ever. In their struggles for survival our common survival is at stake. They are fighting paradigmatic battles for an exodus from a civilization which has no future, either for them or for us others. They demonstrate the importance of an economy of human efficiency. Pushed back into the margins of society, living in fragile eco-systems in remote mountainous areas and tropical forests, they have contributed to the survival of our common planet by *not* exploiting the resources of the mountains, rivers and forests to the maximum, by *not* cutting for maximum profit, by *not* mining whatever would make money, i.e. by preserving the precarious balance which assured water and fertility to the plains. They have been guardians of our future by not opting for 'development'. The dire poverty of many of them today is the result of 'development' processes and should not be confused with the viable self- sufficient economies of earlier days.[3] Against many odds they have kept alive a life-centred logic which we will need for developing a regenerative economy on a new plane.

Significantly, they are no longer fighting rearguard battles, but are coming forward, claiming from modern society the values of freedom and democratic rights which indeed have to be rescued from its crumbling civilization.[4] Their alliances with human rights organizations, eco-movements and others are full of potential for the search of a new civilizational model based on human sufficiency.

Capitalist conquest has not only decimated indigenous peoples. The logic of expansion and accumulation of capital has also attacked the economies of other subsistence peasants all over the world. Stalin's collectivization did the same in the name of large-scale production. Human sufficiency had to make way for economic efficiency. This process is still going on. In India, hundreds of millions of small peasants who have survived till now are facing the onslaught of capital under the new economic policy of globalization, including the new GATT treaty. When they lose their lands, they will have nowhere to go. That is why the resistance of peasant movements against the invasion of MNCs like Cargill and their search for alternative patterns of production with the help of ecologists and critical scientists is of the utmost relevance.

2. Reclaiming the commons

'Free commons have historically been the survival base for rural India and the domain of productivity of women.'[5] Colonial masters, followed by modernizing developers, have undermined the ecological and socio-economic role of common forests and grazing lands. In the process crucial eco-systems were destroyed and with them the survival base of the rural poor, most of all women, the traditional providers of water, fuel and fodder. This process is still going on. But resistance against it is also mounting. The Chipko movement in the Himalayas has become known world-wide for its struggle which projected an alternative approach to the life-sustaining wealth of the forest and its value.[6]

It is often assumed that poor peasants with their needs of land and fodder are a threat to the environment. They are indeed under tremendous pressure, but extensive studies have shown that their survival interests coincide with the protection of the commons against further degradation. Ecologists found to their surprise that only attempts at upgrading deteriorated wastelands were successful in which the rural poor had a stake, a say and a share of the fruits.[7] Both cry out because of exploitation, the earth and the poor. Both together can turn the tide and affirm and enhance life. Both need regenerative production of life and for life instead of destructive production for profit.

3. Re-establishing local circuits of production and consumption

Globalization led by multi-national corporations has not only deprived petty producers of local markets but also consumers of control over the quality and characteristics of the products they buy in the supermarket. Concern over negative health effects of chemicals used in a wide variety of products, eventually combined with concern over the plight of small producers in their unequal competition with giant corporations, has led to various attempts to by-pass the global market and re-establish local circuits of producers and consumers. Health-conscious people discover that modern food-processing and marketing may add monetary 'value' for those who control the global trade, but the nutritional value for the consumer decreases as the distance between 'land and mouth' increases.[8] Initiatives have sprung up to establish direct links between small rural producers, eventually organized in co-operatives, and urban consumers. In Canada and the USA, where giant food companies rule supreme and independent farmers are struggling for survival, a movement for 'community-shared agriculture' is spreading. In it, consumers enter into a contract with local farmers on an annual basis to take a regular share of the farmers' product which will be organically grown and delivered fresh from the farm. Local communities are reviving around such connections, which re-establish people's control over what they produce and consume.[9] In a parallel development, we find people in urban neighbourhoods establishing mutual service organizations in which members barter various services in exchange. All these initiatives indicate that even today, as C. T. Kurien has pointed out, economic activities are not only guided by the calculation of profit.[10] Even in the most modernized societies innumerable transactions, services and exchanges are taking place, directed towards the satisfaction of needs. That is an important base for resistance against the logic of accumulation and for the reorientation of economic life, i.e. of the 'provisioning for livelihood', towards human sufficiency.

III For example: a people's dam

Peasants in drought-prone areas of Maharashtra in Western India, supported by redundant workers from textile mills in Bombay, and by a people's science movement and critical technical experts, have challenged the dominant development model in a way which has inspired movements all over India. So far large-scale irrigation projects had benefitted sugar-cane-growing rich farmers and the sugar mills. Small peasants were left without assured water supply for their traditional food-crops. Unemployment was rampant. Proposals for a more equitable distribution of water, giving priority to less water-consuming food-crops, were rejected by the ruling economic growth lobby. Finally peasants of two villages in a drought-stricken area took an initiative to show that an alternative approach was possible. They planned to build a small dam across a local river, in order to create a water-supply for drinking water, biomass-production, tree-planting, and cultivation of food crops. A society was formed, and all members were to participate in the work and to share the water equitably, including women and the landless. Financing was organized through interest-free loans from sympathizers and through the limited sale of sand from the river bed. It took years of struggle to stop outside contractors extracting sand in an unlimited, ecologically damaging way and to establish people's control over this local resource. They broke the culture of dependency on relief and on 'development'. They demonstrated that by giving priority to the satisfaction of basic needs, all can benefit from common control and sharing of limited resources. They found an answer to the double problem of drought and unemployment.

Meanwhile further experiments are under way, based on the right to water for everybody, considering water as a common property resource, delinked from land ownership. In several places water users' organizations have been formed which regulate the distribution of water on the basis of agreed priorities. New cropping patterns emerge, giving priority to the production of food and biomass for local use, including the production of fibres and wood as raw materials for rural industries. All this requires the participation of local communities in planning and management, whereas new eco-friendly technologies play a supportive role. At the same time political battles have to be fought to secure space for these experiments. In order to resist temptations to get hooked up to the dominant development model, the cultural factor plays a crucial role. Recalling a long tradition of popular anti-brahminical mass struggles, the people's dam is named after the legendary peasant king Bali, whose just rule will return in spite of all efforts to destroy him.

IV Features of an emerging new paradigm

Our expansionist, fossil-fuel-based capital and/or state-centred industrial civilization has reached its limits. Those who have ears hear the cry of the earth and of the poor. Others may wake up under the impact of ecological catastrophes and explosive social and political disruptions. Capital struggle for accumulation in the face of deepening crises may take ever more destructive forms, unless enough people are ready to move towards an alternative type of civilization. Such readiness will grow if experiments in that direction are showing the way. What already emerges is the relevance of a primarily biomass-based, ecologically sound, decentralized, need-orientated, locally diversified, people-controlled system of economic livelihood, centred on the production of life and for life, embedded in communitarian socio-cultural structures nurturing the satisfaction of diverse socio-cultural needs, and protected by political structures which provide space for people's participation and control.

(*a*) The laws of entropy imply that no new energy can be created, while unavoidably part of the non-renewable energy we spend turns into unusable waste. Solar energy offers a viable way to increase the production, not of all that industrial society has been producing, but of those goods which are needed to feed, clothe, shelter and provide for the dignified life of a growing global population.

(*b*) One of the basic principles of an ecologically viable economy is to start from using – as much as possible – local resources for the satisfaction of local needs. That was Gandhi's concept of *swadeshi*. It does not imply the ideal of autarchy. But it is a reversal of the unsustainable export-focussed global division of labour. Exchange and trade would once more play a supplementary role, to provide for goods which cannot be produced locally.

(*c*) At present global companies try to homogenize human 'needs' through gigantic, brain-washing advertising campaigns, in order to make maximal profits in global markets. This is both ecologically non-viable and culturally destructive. This cultural imperialism of Western consumerism has to be countered by a creative re-affirmation of culturally and ecologically diversified needs.[11] Such a resistance and reorientation requires a cultural revolution, carried forward by revitalized communities based on the reorganization of the production of life.

(*d*) People's control of resources has become one of the key demands in the struggle of peasants, fishing communities and others. Centralized administrative and bureaucratic structures, be it of the World Bank or of state planning boards, are unable to rebuild devastated environments at the base. Only local communities can do this, as it requires the involvement of people who do the actual work of nursing, caring and nurturing nature back to health. Their motivation cannot be sustained by cash alone. It depends on their participation in organizations of the local community and on their being assured of a share in the fruit of their labour, in the supply of water, fodder, building materials or whatsoever. Of course, on other levels, from the regional to the global, representative political and administrative structures are needed to counter the forces of destruction.

(*e*) Women have been doing and are doing the work of bearing children, nurturing and caring, providing water and fodder, preparing food, often also growing it and performing other life-centred tasks. Men and the capitalist system at large have been exploiting this unpaid labour. Parts of the women's movement have looked for liberation from their burden in the direction of the mechanization of household labour. A more viable alternative is envisaged by those who advocate and anticipate a new affirmation of the value of nurturing/caring work, orientated on the production of life and for life, to be shared by women and men alike.

(*f*) Capitalism's philosophy of utilitarianism considers labour as pain to be avoided and leisure as pleasure to be sought. Physical effort has to be avoided except in sport. Motorization and mechanization in all areas of life are being justified in that way. The ecological and psychological damage which this causes challenges humankind to rediscover the dignity and value of regenerative constructive and productive work in interaction with nature, and of the social satisfaction it can give if it is an integral part of commonly planned and shared tasks. Eco-friendly and people-friendly technological innovations will lighten the toilsome part and praxis-orientated upgrading of skills will increase the satisfaction which common involvement in production for life is able to give.

V Struggle and creation

The history of experiments in search of an alternative seems to vindicate Karl Marx's warning against utopianism and reformism. They tend either to remain isolated set-ups for a few people who want to opt out, or, if they are successful, to get hooked up and absorbed by the expanding capitalist system. Arguing against the Marxian critique, Martin Buber pleaded in the 1940s for a fresh review of the utopian socialist traditions, pointing especially to the experiment of the kibbutzim of Jewish settlers in Palestine.[12] However, these socialist communes got subordinated to the military and political exigencies of the state of Israel, and seem to have lost their function as signposts to a non-capitalist society.

The other world-historical experiment of the Soviet Union, based on the conquest of state-power as the precondition for all transformation, has also failed. One of the reasons in my opinion was that it tried to emulate the old capitalist paradigm of unlimited growth based on large-scale industrialization.[13] This happened at the cost of the peasantry, of the environment and of democratic participation. Maybe more 'utopian', community-centred experiments could have shown a more viable way.

Today awareness is growing, and signs are pointing in the direction of a feminist eco-socialist paradigm. But the maturing of political breakthroughs may depend on the tenacity and practical imagination of marginalized people, pioneering visionaries and their supporters in demonstrating the practicability and attractiveness of an alternative paradigm, while increasing eco-problems and economic breakdowns will force more and more people to look for a way out.

In this we face the basic problem of sustaining voluntary bonding and communitarian commitment. Pre-capitalist societies were held together by tight social control based on the power of tradition. Post-capitalist societies cannot be held together in the long run by a dictatorial state and moral-ideological propaganda. What will be needed is a covenantal politics on all levels in which personal choice and commitment are sustained by participatory organizational structures and by a multi-faceted cultural praxis, drawing on traditions of the past as well as on visions of the future, organically related to daily life needs and desires in the present.

Theologically it is a matter of affirming simultaneously a commitment to God's covenant with the earth and to God's covenant with the people on their way from slavery to the promised land. The struggle for rights and the nurturing of life have to go together. The cycle of seasons and the historical project of hope, the struggle for a new society, are not opposites. The struggle aims at creating space for all to sit in peace under their fig trees. However, for that, we have now already to preserve the seeds and prepare the ground and inspire people, while fighting against the powers that threaten to turn the earth into a desert denying life to its inhabitants. As India's great worker and peasant leader, the martyred Shanka Guha Nyogi, put it: we need creation and struggle, struggle and creation.

Notes

1. See the discussions in: *Capitalism, Nature, Socialism. A Journal of Socialist Ecology*, New York; E. Altvater, *The Future of the Market*, London 1993; R. Williams, *Resources of Hope. Culture Democracy, Socialism*, London 1989.
2. K. Marx and F. Engels, 'Manifesto of the Communist Party', in *Selected Works*, Vol. 1, 111.
3. See R. Goff, *Land Without Evil*, London 1993.
4. See *I. Rigoberta Menchu. An Indian Woman in Guatemala*, London 1984, and cf. the reports on the uprising in Chiapas/Mexico, January 1994.
5. Vandana Shiva, *Staying Alive. Women, Ecology and Survival in India*, New Delhi 1988, 83.
6. Ibid.
7. Anil Agarwal and Sunita Narain, *Towards Green Villages. A Strategy of Environmentally Sound and Participatory Rural Development*, New Delhi 1989.
8. See Brewster Kneen, *From Land to Mouth. Understanding the Food System*, Toronto 1989.
9. See *Ram's Horn*, published by Brewster Kneen from Toronto.
10. C. T. Kurien, *The Economy. An Interpretative Introduction*, New Delhi 1992.
11. See the last article by Fr S. Kappen, written shortly before his death in November 1993: 'Towards an Alternative Cultural Paradigm of Development', in *Lokayan Bulletin* 10.4, January 1994.
12. Martin Buber, *Paths in Utopia*, Boston 1958.

Ecology from the Viewpoint of the Poor

Eduardo Gudynas

The link between extreme poverty and the environmental crisis is an urgent matter. In Latin America about 44% of the population lives in poverty. This represents about 183 million people; 21% live in conditions of extreme poverty. That means that most of these people are unemployed or under-employed; more than half of them have inadequate housing; infant mortality rates are highest among their children; and of those children that survive, 15% do not finish their schooling.

The poorest people also suffer directly from environmental problems. In rural areas, where 61% of them live, they suffer from loss of land, soil erosion through bad farming, abuse of agro-chemicals or deforestation. The other 36% are nearly all crammed into big city slums, where they have to cope with an unhealthy and dangerous environment. Because of this close relationship between the environment and poverty, we need a social ecology.

The revival of ecological vision

This connection between poverty and ecology is not new. The concept of marginality, which is central to this argument, originated from human ecologists. Robert Park, a sociologist in the University of Chicago, began referring to 'marginal people', not with the present-day meaning, but as 'hybrids' between two cultures. 'They are people on the margin of two cultures and two societies, which never fuse or completely interpenetrate.' He derived the concept from his studies of immigrants who came to the city of Chicago and did not adapt completely to the local culture. The non-adaptation could be positive.

Over the years, the concept changed, as new studies and concerns were added to it. In Latin America many analyses were made, even though here the ecological viewpoint receded and the stress was on spatial marginalization, referring to those who lived on the 'edges' of cities.

Today the ecological dimension is returning to the fore. First, there is a recognition of the enormous ecological crisis from which the continent is suffering. This includes the high rate at which animal and vegetable species are continually becoming extinct, the disappearance of some separate eco-systems, and vast pollution damage both in country and town.

Together with this environmental crisis, we are becoming aware of the emergence of the 'new poor' living in conditions of extreme marginalization or exclusion. These words describe a new type of poverty, which has arisen in recent years, and is more violent, excluding and segregating. A few years ago in Argentina Maria del Carmen Feijoo rightly denounced the fact that: 'Historically speaking, true poverty is arriving now: social groups who fall into it, new poor side by side with the traditional poor, who make up a new panorama of poverty and disorientation.' She adds that this new poverty generates the serious social problem of 'exclusion as a system of social recognition'.

Whereas in the past there was an attempt to integrate and restore the most impoverished sectors, today they are excluded and marginalized. No attempt is made to offer them more equality; instead they are regarded as different. Instead of help to improve their lot, they may be given alms. Whereas in the past there were denunciations of and rebellions against the 'exploitation' of the poorest, who might have to work exhaustingly long hours, today, for the excluded who have no work, becoming 'exploited' in the old sense would be a privilege. Of course this is a grave injustice. Many have noticed the growth of an 'us' and a 'them', in which the excluded are marginalized at all levels, including the ecological level.

The viewpoint of social ecology

It is the connection between the social crisis and ecology that has led to the birth of social ecology in Latin America. Unlike what has happened in North America or Europe, in Latin America this movement has been practical. Reflection is combined with action, inspired by an ethical commitment in defence of life.

It has explicit practical objectives in defence of life for people, plants and animals. Social ecology recognizes the close and continual interrelationship between human and environmental systems. Neither can be separated from the other. This perspective has also been adopted by committed Christians, to help them understand and serve the very poor, while at the same time conserving the environment.

Social ecology is also a reaction against an anthropocentric and dominative philosophy. Instead it offers a utopia. There are seeds of change in our Western culture, to which indigenous Latin American cultures, ethnic minorities and peasants are making an important contribution.

The utopia they have in mind is a re-encounter between human beings and nature, and among human beings themselves. It is a utopia which unmasks the current ideology, shows its limitations and contradictions, and points towards a possible future. It is a mobilizing utopia.

Table 1. Central Tenets of Social Ecology

(*a*) Human beings interact intensely and continually with the environment. Neither can be studied in isolation from the other, because each determines aspects of the structure and functioning of the other.
(*b*) The interaction between human and environmental systems is dynamic and develops in time and space.
(*c*) The definition of the environment is dependent upon how the human system defines itself.
(*d*) The environment is complex and heterogeneous in time and space.

The environment of poverty

People are excluded economically, culturally, politically, and also ecologically. The same causes that marginalize them from the economic process, cultural and political life, also drive them into inappropriate environments. This is the environment of poverty, where social poverty coincides with ecological poverty.

The ecological exclusion suffered by the poorest has various characteristics. First, the excluded sectors are driven to live in the worst environments. Geography or physical conditions make them inadequate for agriculture or urban living. They are areas where flooding occurs, or landslides or storms. There is a high risk of catastrophe, which may kill or cause disastrous losses. In other cases, the poor are housed near factories, which heavily pollute the environment. There is no access to drinking water, or if there is, the water is contaminated. There is usually a high level of other pollution (especially of the air through waste emissions) and a high incidence of contagious diseases.

Secondly, the impoverished sectors occupy these sorts of environment because they are 'free'. They are free because of their bad environmental quality. Generally, the settlements house many families with a high population density. Infrastructural and social services are non-existent or bad. Ecological exclusion (living where no one else wants to) reinforces social exclusion.

Thirdly, because they are environmentally and socially excluded, these groups can only maintain precarious relations with the formal production system. Most are involved in the informal sector, in work which is casual and unreliable, without legal protection, and which brings in very low wages. The rest of society allows them access to its waste products as a slim chance to survive.

Fourthly, these groups have developed their own techniques to resolve some of their own problems. An obvious example is self-built housing using discarded materials. Likewise, they have developed cultural patterns for various forms of public and private behaviour, relationships of strong solidarity, and rules for the use of common resources and self-defence.

These processes have various consequences. In cities the poorest human groups live on the periphery; many live close to roads entering the city. Or they take advantage of intersections with bridges and streams. Others invade urban waste grounds and build slum settlements, such as the *favelas* in Brazil, the *campamentos* in Chile, the Peruvian *corralones*, the *callampas* of Colombia, the Argentinian *villa miseria* or the *cantegriles* of Uruguay. Their shacks are very precarious, built of cardboard, plastic or metal. There is no sanitation and contamination everywhere. There are high levels of infectious diseases and minimal public services.

Poor people also suffer exclusion, with marked ecological effects, in their access to work. For many of them, getting a more or less steady job is a dream, and they often end up doing work which no one else wants to do. They take jobs in dreadful environmental conditions, sometimes with grave risks to health. They breathe in toxic gases, and are exposed to deafening noise or the risk of accidents.

In the country, extreme poverty leads to survival strategies which have a very negative impact on natural resources. Some people are driven to poaching, both to feed themselves and to obtain a little money to help them out. Others exhaust their soil in order to produce enough to eat. Living on bad land, lacking water, without additional resources to improve their crops or livestock, they end up causing more environmental damage, which in turn increases their poverty.

Poverty and environmental damage

Poverty and environmental problems are closely linked. There are two versions of the way this link works. On the one hand, there are those who hold that the poorest sectors are the ones which do most damage to natural resources. Others do not agree that the poor have the greatest responsibility for environmental damage; they are merely the ones most affected by it.

The first position is based on the 'coincidence' between excluded groups and certain environmental problems. For example, the poorest peasants appear to be the ones most likely to over-exploit their land, leading to soil exhaustion, or to the cycle of tree-felling and cultivation, which is a major cause of deforestation in a number of Latin American countries. They conclude that it is the poor who destroy most natural resources.

The second position derives from an analysis that tries to go beyond appearances. It shows that both social exclusion and environmental deterioration result from a certain type of development. This development model leads to excessive consumption and opulence, which in turn causes the exclusion and marginalization of many people.

It is true that in many cases poverty itself reduces opportunities to use natural resources in a way that would take more care of their conservation. In many cases a vicious circle forms, because not conserving natural resources increases poverty. In general, we find that the decreased fertility of the soil, problems of access to water, and difficulties with regard to other natural resources mean that rural areas lose their productivity levels and the risk of natural catastrophes increases. Then the rural poor earn even less or suffer disasters, and never manage to get out of their poverty.

As S. M. Mohamed Idris of the Third World Net (Malaysia) says, ecological concern is not a luxury for the poor countries of the South, and 'contrary to what the elites maintain, the poor are not the cause but the victims of the destruction of the environment'.

On the other hand, many impoverished sectors employ very ecological production processes, which respect natural resources and recycle waste. So there is no causal link whereby poverty necessarily creates environmental damage. Both social poverty and environmental destruction are really symptoms of a deeper problem: a type of exploitative development, obsessed with efficiency and maximization, which manipulates and dominates both human beings and nature.

Strategies for ecological survival

It is a remarkable fact that, despite their poverty, the excluded sectors have managed to establish survival strategies which are profoundly ecological. One of the best examples is the sorting and recycling of waste practised by many groups.

These activities are increasingly common in Latin American cities. The 'tinkers' of Buenos Aires, the 'thrifties' of Lima and the 'scavengers' of Uruguay gather rubbish, refuse and waste-products from the city, sort them out, and refurbish those that can be reused.

This is an ecological activity, because it does not use new natural resources, and because it encourages recycling and thus reduces demand upon nature. It also reduces the generation of waste, which is one of the big environmental problems facing modern cities.

We should not forget that this work is done in conditions of poverty and marginalization. The reclaimers have to walk for many hours, often at night – in cold or rain, sometimes pushing their carts. They sort the rubbish without protection, often risking their health. Nor should we forget that these waste-sorters are not formally integrated into the productive process, and can only make use of its final results: rubbish.

Nevertheless, their activity is an important lesson to us. This rubbish is not wasted but put to use. It also shows that even in poverty, sustainable ecological practices can be generated. However, this potential should *not* be used as a reason to maintain that the role of the poor is to recycle the rubbish of the rich.

A closer examination reveals a deep respect for life, close relationships of solidarity and a popular religious feeling in tune with the gospel message. All these add force to what those living in environmental and social poverty are telling us.

The crisis of justice has an ecological face

Clearly both poverty and environmental damage are above all a consequence of a particular view of society and the world we live in.

As the Mexican researcher Exequiel Ezcurra said recently: 'By placing growth before distributive justice – the eternal dilemma of human societies – in Latin America we are seeing the development of a high-spending and consuming sector of the population, side by side with the material degradation of a marginalized sector, whose hopes and attainments of previous decades have come to nothing. They are forced to try to survive by scavenging from their surroundings and on the leftovers of the more consumerist sectors.'

Thus both humans and the environment are excluded. It is forgotten that our natural surroundings contain animal and plant life and these are reduced to a mere collection of natural resources. Likewise men and women in all their diversity are cast aside and become mere human resources. These resources are used interchangeably and pitilessly pressed into service, till they are no longer useful.

We should not therefore be surprised at what is happening in some Latin American countries. For example, excluded people in rural areas are occupying marginal lands, where they try to survive and end up with problems such as erosion or deforestation. Their options are limited. The important thing for them is daily survival. Once the chain has been formed, poverty generates more environmental damage, which in turn keeps people in poverty. Poverty impels them to it.

The other face of this system is opulence. Wealth and resources are accumulated by excluding other people's access to these resources. High levels of consumption are not only the principal cause of environmental deterioration, but also generate these mechanisms of exclusion.

The best environmental policy is to eradicate poverty

It is impossible to solve our present environmental problems without solving the problems of poverty and exclusion.

It is urgent to solve the environmental problem because it affects people's very survival. We do not deny the importance of global issues, such as the loss of the ozone layer or global warming, but these must not distract us from our own problems. Bob Sutcliffe is right when he warns us that 'most of the population of the Third World today are going to die from more immediate causes than climate change through excess of carbon. The solution for them has to be found in a radical change in international relations, to put an end to the pressures forcing them to transfer a large part of what they produce to the rich countries.' We can go further and recognize that a solution to this problem will overcome the serious crisis in our culture.

Social ecology and theology of life

Western culture has a metaphysics for our conception of nature. We see nature in terms of our perceptions, values and inherited ideas. Little by little, both modernism and postmodernism have divested nature of any transcendence, acknowledging the vital web by which ecosystems live. This does not mean we now have no metaphysics. Instead of the metaphysics we had, we now have a new metaphysics, which leaves life itself without any meaning. Everything becomes just resources; both human beings and the environment become mere human or natural capital. The destruction of forests or the exploitation of workers are new forms of efficiency in the management of this capital.

The fundamental question is whether we cease to value nature when we try to dominate it. C. Lewis raised this question nearly thirty years ago when he denounced the fact that 'we are always conquering nature, because nature is the name we have for what we have conquered . . . '

Some recent ecological fashions do not necessarily represent a better option. They are mere fashions and buzz words, that do not result in committed action. They suggest that wearing a T-shirt printed with the slogan 'Save the Forests' will effectively save the forests. They simply act as a social anaesthetic, because they distract from the search for committed options leading to action for change. Nor is an isolated commitment to save particular plants or animals sufficient, because it ignores the human drama being played out around us.

The challenge is particularly clear and acute for Christians. Their struggle for the dispossessed must be a struggle for the environment, and the preservation of nature must also be a social struggle.

Profound changes are necesary in our vision of the cosmos. We need a new metaphysics of nature, which overcomes the dualism separating us from it. It would enable us to recover the mystery and magnificence of being part of nature, part of life. This is the most important contribution of social ecology to a theology that is sensitive to the problems of the environment and poverty. Theology cannot become sensitive on its own. When it is out of touch, it becomes a theology of domination or despair. True theology, born from an encounter with God, must be a commitment to life. We have been creating a theology of liberation; now is the time to broaden it into a theology of life. Many base groups, both in cities and in the country, are working on this now and not waiting for academics to do it. Let us hope that on this occasion we do not get there too late, too late for humans, and too late for nature.

Translated by Dinah Livingstone

Bibliography

H Assman, *Clamor dos pobres e 'racionalidade economica'*, Sao Paulo 1990

L. Boff, 'La espiritualidad franciscana frente al desafío del desequilibrio ecológico', *Vida Espiritual* 50, 1976, 50–61

E. Gudynas, 'The Search for an Ethics of Sustainable Development in Latin America', in J. R. Engel and J. G. Engel (eds.), *Ethics of Environment and Development*, London 1990, 139–149

E. Gudynas and G. Evia, *Ecología social. Manual de metodologías para educadores populares*, Madrid 1993

J. Ramos Regidor, 'Hacia una reconversión socio-ecológica de la sociedad', *Crisis, ecología y justicia social*, Cuadernos de Peregrinos 2, Montevideo 1991, 63–87

F. Tudela, *Desarrollo y medio ambiente en América Latina y el Caribe. Una visión evolutiva*, Madrid 1990

Principles for a Socio-Environmental Ethic: The Relationship Between Earth and Life

Charles Richard Hensman

The principles for a socio-environmental ethic discussed here will be those arising in the course of: first, the overcoming of what the physicist David Bohm describes as 'an overall fragmentation in our general attitude to reality';[1] second, the close interconnections between the newly empowered, life-appropriating social practice of *the people*, all over the world, and the renewal of the living Earth; third, their having co-operatively to create new, non-exploitative, non-hierarchial, self-managed and life-enhancing technologies and social organizations instead of trying to recover or recreate for the future an idealized pre-capitalist past; fourth, the firm and rapid, but compassionate, management of the revolutionary and comprehensive transformation of present systems and conceptions of humans as super-powers, ruling classes and lords – to create conditions equally fulfilling, contemporary, flourishing and peaceful for humanity and Earth, whether in Eritrea or Manhattan, Shenyang or Hamburg, Mexico City or Borneo, the Dalits or the Gabra.

Overcoming fragmentation

Thirty years ago, the Prime Ministers of Sri Lanka (then called Ceylon) and of India met in New Delhi and made a deal over the fate of a million or so persons from working-class families in the hill-country tea plantations who had earlier been disfranchised and made 'stateless'. Like the longer-settled Sri Lankan Tamils and Moors in the north, east and other parts of the country, these fifth-, fourth- or third-generation descendants of poor South Indians who had been brought in colonial times by the British to set up the coffee and tea plantations in the central highlands of Sri Lanka spoke Tamil.

During the following years the forcible uprooting and shipment of tens of thousands of families was organized. A number of us, at that time not endangered, protested against this unjust and inhuman treatment of fellow Sri Lankans, but to no avail. The victims' grief and outrage took many forms. One of them expressed in poetry the sadness of being torn away from his much-loved and indescribably beautiful homeland. A barely literate worker's 'love' of the incredible beauty and peace of what nature's prodigality and skilful art had created over millennia was not something in the calculation of ambitious politicians. Sri Lanka was well on the road to having one-tenth of its population as 'displaced persons' in the 1980s.[2]

Around the plantations, in which extended worker-families had to live in dingy 'line-rooms' 10 feet by 15 feet wide, were roads and a railway from which could be seen the panorama of more or less extensive forest-covered mountains, hills and valleys, interspersed with small, terraced farm plots and the slopes of neat, shrub-covered tea plantations, all lush, of various greens, and teeming with life. The bonds that were forcibly being severed for the plantation families were only partly those with their places of residence and labour. They were bonds with the wider community – with those whom they met at festivals and at political and trade union meetings and with Sinhalese villagers. They were also bonds with the sources of life itself. Miserably paid, and with much less schooling and health care than were their rightful share, many of them could at least commune with Mother Earth at her most beautiful. In the ancient *Atharva Veda* there is that wonderful poem *Bhumi Sukta*, known as 'Hymn to the Earth', and who knows how Earth herself was hurt by the forcible expulsion of her children?[3]

'Up-country' Sri Lanka – the 'artistic' perfection of it and its fertility – was the work of nature and humans over tens of millennia. At that time, thirty years ago, the forests, with their varied plant, insect, bird and animal life, were more 'alive' than they are now. The brightness of the sun, reflecting the different greens and blues and pinks and whites, was purer. There the air, rich soil and clean water, freshened and enriched continually by the eco-system, was most life-restoring. The quiet and undisturbed wilderness of those mountain tops could also bring enlightenment, spiritual refreshment and the call to a new beginning.[4]

Like their fellow Sri Lankans who protested then, there are considerable numbers of people who perceive that many things going on around them are inhuman and unholy: in the corruption of politics and of international relations; the heartlessness and cynicism of groups and nations, including their own, which follow the top profit-makers in advertising positively harmful commodities as 'what the public wants'; the reckless destruction of tropical forests; the way science is done; the way women and girls are humiliated and hurt; the way millions are being condemned to live, within the sight of billionaires, in dark, airless, waterless and filthy 'bastis'; discrimination against minorities; the trickery of 'structural adjustment programmes' and GATT reforms; religious fundamentalisms of different kinds and the failure to take religion seriously enough. Are we talking about devising in the economic interest more efficient methods of controlling dangerous exhaust emissions from the million vehicles which clog the highways and thus keep the automobile and oil companies at 'the top'? Are we seeking economical ways to discover how to reduce the poisons used to control plant pests and to increase yields? Each of these turns out to be one of the massive and ongoing 'problems' the system continually breeds.

There is valuable work being done by some groups and people's movements to correct what is wrong. But it is done more than ever before in the form of fragmentary attempts to define and tackle 'problems', using standard categories. The attempts to tackle these, even in global summit meetings of a few non-governmental organizations and governments, rarely undermine the 'infrastructure' of liberal assumptions. They amount too often to a fragmented approach by individualistic-minded and fragmented groups dealing haphazardly with political-economic-cultural contradictions and unities. The global powers-that-be may even fund such non-governmental organization activity, as it helps people to live with their frustrations and deprivations. The inclusive problem of how humans could become creative so that social systems, genders, economic formations, nations, classes and castes relate differently to one another and to Earth as a living system requires a different kind of agenda.[5]

Really ending the enslavement and abuse of humans in order to end humankind's alienation from nature

There has been a worldwide tendency among some groups to agitate and act over human-caused destruction of the environment, while ignoring the desperate struggles against human-caused blighting of most of humanity. There has been another tendency to regard human life-activities as the main threat to earthlife, indeed to regard the satisfaction of everybody's human needs as the main problem of ecology. Needless to say, this is a 'Northern' and Europocentric view of humankind's identity and capabilities as well as of what a post-capitalist technology and life-style can become. It does not face up to the questions of at what cost, by whom and for whom, and by whose authority the world should be reorganized and managed.

There is more to be said than space allows about the beauty/goodness of the earthlife that came and still is coming into being, in response to the Creator's word (Gen. 1.11–25); and also about Yahweh's assurance, given through Moses: 'The whole world is mine' (Exod. 19.5; also 9.29). It is not Pharaoh's. It is not capital's. All creation is in the economy-polity of the new order. We have only to note the range of references in Luke 6.20–49; 12.1–34, 54–59 to 'real life'. In the more or less 40,000 years of human life-activity, humans, in combination with nature, have laboured with their 'natural' 'potentiality', 'creativity', 'vitality itself', as Karl Marx put it, to transform and diversify the material conditions of life.[6] It is the kind of 'growth and development' that makes us continually dependent for our capacity to work on classes of fellow-humans who have usurped as private property bits and pieces of the fruits both of earthlife and of human creativity which threatens creation.

Humans are history-making creatures. The old history, which turned out to be a story of violence, pride, greed and arrogance, is still going on. More than ten thousand years ago, a few began to lord it over earth's resources and to monopolize them, to dominate over women, and to put others to work for them. They arrogated to themselves the power and the right to rule, to work to death both subjugated peoples and animals, to kill as they pleased. That complex history, which includes the ancient Egyptian, Assyrian and Roman empires and the modern British, French and US American ones, is what concerns us at present. If that is the only history that there can be, if the overconsuming, profit-obsessed, lawless, universally interfering, independence-denying superpowers are indeed 'the chosen people', models of true human success, then the vast majority as well as earth cannot escape subjection to death-dealing processes. A kind of 'holiness' can be attained only through 'unworldliness' and passivity. But there *is* another history, which is sustained by faith in a promise, by hope and by love, with a mandate, rationality and values which are not those of even the most 'enlightened' imperialism.

The other history being made is that being made by *the people* as a class – those who were or are without land to live on and play on, or to work with, or to get what is necessary to work and to keep alive; those whose vitality and creativity have been suppressed; those without the water, the air, without the light that we have already mentioned; those, especially the women, whose bodies are not theirs, and who do not have the space to move about in which the poorest of their menfolk have; those whose working day fully exhausts all their vitality and all their time. So the story of liberation – in which the restoration of the land to God's creatures will begin, one within reach of God's care, one in which emperors, landlords and company presidents will be judged, and much else – is the new history. It is only in a world order in which the hundreds of millions who are actually starving, those without anywhere to live, the tens of millions kept unemployed in order to maintain profits, those who collectively are completely subject to others, are all restored to dignity, health and human responsibility, when war-mongering comes to an end, that care of earth-life will be high on the agenda. It is not the eradication of hunger, unemployment and poverty that will destroy Gaia, but rather the continuation of capitalism.

Those who, here, there and eventually everywhere are given back the gift of land and their vitality will have to learn how to come to live humanly. God has promised to show them how – whether they are 'Buddhists', 'Hindus', 'Muslims' or 'Christians'. They will do so by experiments and experience. But they will certainly be hated and maligned by the lords of the old dispensation. No doubt they will learn to live frugally, to conserve water, to invent people-controlled technologies, to form co-operatives as they did at one time in China, and to disperse power. For vast urbanized populations, where renting, buying and selling of 'real estate' can be highly profitable, for the poor to reconnect with the sources of clean water, fresh air, sunlight and food, and housing, too, will be an enormously difficult task. They will have to learn how to get to Jerusalem. Their prosperity and peace, if they are faithful and persistent, cannot worsen the threat to the environment.[7]

Becoming self-reliant creators of a post 'Vasco da Gama era'

The 'Vasco da Gama era', the end of which the Indian historian K. M. Panikkar optimistically expected in the 1950s, will soon be 500 years old. We in Sri Lanka entered it in 1505, just as our neighbours in Malacca did in 1511. What Western historians incorrectly describe as the 'voyages of exploration and discovery' were mainly for the purpose of conquering, occupying, annexing and exploiting as much of Earth and as many peoples as could be defeated for gain. For us in this part of the Third World, as in Latin America recently, there is the awareness that the colonialist and imperialist forms of contact among peoples have, while being beneficial in a few ways, resulted in the looting and destruction of material-cultural productions of tens of thousands of years, the distortion or destruction of delicately constructed forms of life-activity, the loss of complex and richly meaningful vocabularies, the disappearance of hundreds of languages, and the loss of several species and of peoples. What is being explored, discovered and studied now about biological diversity or the ethnic problems in our lands by the present-day successors of the Portuguese invaders is part of a process of subjugation and impoverishment which has to be ended for good.

Some people's movements to stop massively conceived 'development' projects which would cause more harm than good (as the protesters argue) are world-famous. At another level, polluting industries are being restrained, polluted waterways are being cleaned, new technologies are being devised: these are promoted and advertised as 'people-friendly'. This is not the place to recount the stories. In a discussion of socio-ethical principles, what we need to ask at this point is how feasible are the various attempts, however heroic, to reform and remedy, reorganize and legislate, so that Earth can be conserved for 'sustained' exploitation.[8]

In Asia, particularly in South Asia, more than perhaps elsewhere, there has been an influential tendency to conceive of the principles for a socio-environmental ethic as a total rejection of 'development' and modernization of the kind experienced in European and European-influenced societies and cultures. Not only the destruction of the environment but other disastrous tendencies (e.g. in health care) can, it is argued, be halted and reversed if we go back to the past and start again. There are powerful voices to be heard in this cause, and very distinguished ones, too. There are similar arguments by less renowned persons than Vandana Shiva in India.[9] Because of the long, very complex and still inadequately researched history of the peoples of South Asia, let alone the rest of this populous continent, discussion of this topic will be very controversial. However, in the context of our present discussion we must note these questions, at least. Can admirers of their own ancient national or tribal traditions set as their goal the creation of an imagined or real 'ideal' pre-industrial society while most of humankind, four billion of the poor, decide how to go forward into a twenty-first century more advanced than that envisaged by the top companies and banks, racists, chauvinists and Christian and other fundamentalists? Can we truly create an alternative order of mutual learning and mutual respect, which includes all of life? Can we separate the critiques of 'development' from a total agenda which includes worldwide networks of action from below by the people of the world, the disfranchised, our sisters and brothers who have been written off as good as dead even as they are born?

All peoples have a variety of creative abilities. In Sri Lanka there was iron production as early as the ninth century BCE. From the first century CE the unnamed mass of the Sri Lankan people devised and built up remarkable systems of water conservation, distribution and use. All over the island they got to understand the richness and diversity of earth-life and developed it. Of course, millions of years (if we accept the standard geological chronology) of nature's creative power in this part of earth was responsible for the fact that, of all the countries in Asia, Sri Lanka has the greatest variety of eco-systems and the most genetic diversity. But the products of human ingenuity, in co-operation with Nature, must also have been significant. To take an example of cultivars (strains and varieties devised by humans), of the 120,000 varieties of cultivars in rice-production known in the world, 2,800 were developed in this small island. Given their natural potentiality, 'ordinary' women, men and children today, closely related to and knowledgeable about particular terrains, weather patterns, techniques of production, and so on, can still draw on, transform and enrich the life emanating from Earth.[10]

The subjection, under the rule of capital, of more and more economies to market forces, has been glorified as the triumph of economic rationality. Whatever Earth gives life to or humans have produced and organized is evaluated as only a 'resource' in parts (e.g., elephant tusks) or as a whole for profit-making manufacture or commerce. What for generations, centuries or millennia it has been part of – a species, physical or spiritual or emotional nourishment, an eco-system – begins to disintegrate and to disappear. The very notion of bits and pieces of what nature may have taken millions of years to create being traded or used as 'resources' to make useless things comes out of the notions of private property and a US-dominated world-market.

The ecological changes brought about by the commodification of nature are eroding the bases of Sri Lanka's ecological diversity. The various types of forests, for example, are not only the habitat of precious crop plants, trees, livestock and small organisms. They protect Earth's richly life-giving ground surface (the soil with its micro-organisms), they regulate the water supply, affect the climate, and support a host of economic activities: the domestication, cultivation or gathering in the wild state of fruits and nuts, herbs, vegetables, spices, medicinal plants, bamboo, rattan, leather and fibre; they support a wide variety of local industry and meet day-to-day consumption needs. They preserve precious genetic resources as yet undiscovered. But there is more that is at the easy disposal of the 'purchasing power' of the very rich: the various types of grasslands, and the mangroves, mudflats, streams and ponds.

Deforestation for the sake of serving foreign markets and local entrepreneurs, monoculture, 'structural adjustment programmes' and big 'development projects' which politicians find highly remunerative is death-dealing activity. It used to be said that at the beginning of the century tropical forests covered about 70 per cent of the island. The first scientific survey, in 1956, showed that the forest cover had declined to 44 per cent, about 2.87 million hectares. By 1992 only 20.2 per cent of tropical forest (1.33 million ha.), was left, with the current rate of decline being 54 ha. a year. It is because independently and democratically worked out *Sri Lankan* solutions to the land problem, violence and massive corruption were urgent that the mass of the people voted twice as they did in the past year.

Life-giving life has to be a just and thoroughgoing reordering of the whole world order

Does self-reliance mean that people's organizations in different countries have to work in isolation? Does it mean that in each of our countries or regions (e.g., Central America, or the Maghreb, or Southeast Asia) have to co-operate only among themselves? The answers to these questions must take us some way to becoming clear whether the end of death-dealing ways of treating humans (as genders, as races, as classes) can be brought about by local reforms. If it cannot, do we not have to struggle and work for a fundamental and global transformation, focussed on North America, Japan and Western Europe, of the market-dominated and imperialist 'one world' which has a stranglehold on the people and on Earth?[11]

As we get down to 'basics': billions of 'ordinary people' in hundreds of thousands, perhaps millions, of local communities and environments, can stop co-operating with the death-dealing makers of the old history. Our home community is where we start from, and home is where we return to. In it and around it is where daily and for a lifetime acts of love and service are performed, where celebration goes on, and the joys of life are most often experienced. It cannot be from where some strong-armed man can eject us, or into which effluent is emptied or garbage is dumped. It should not be where one leaves hungry and tired for the long journey to 'work' and to which, extremely fatigued by the journeys and by toil done under orders, one returns with hardly anything in one's purse or pocket. We recall where this article began: the brightness reflected off the greenery and the flowers; the fresh and bracing air created by Earth; the sparkling and unpolluted water to drink, bathe and cool oneself with; the space open to the clear sky, and where children and adults can play, dance and talk. We can add: the relaxed, secure, usefully good and peaceful living in harmony with other communities and Nature.[12]

If I looked for a market to get this, where would I find it? Like millions of others I may not mind wearing the same homewashed blouse, shirt, skirt, pair of trousers or sarong for ten years, and using the same saucepans and dishes, even if they are a little battered or chipped. The 'life' gaudily advertised over and over again on television and hoardings does not bring tasty and nourishing food and drink, joy, contentment, friendship, peace. In it the very best that one can do or give is, for most of humankind, not wanted. One has to produce the life one wants.

Many of the things my neighbours and I need and want we can produce co-operatively, and with little expenditure of time, energy and materials. The motor vehicles and the oil consumption can be cut down by 75 per cent. New machines will have to be designed, and housing reorganized. The designers, machine operators, craftspersons, fisher-people, housebuilders, story-tellers, teachers, writers, town builders, transporters, and so on: who else are they but we? And the land, with its fertility, whose else is it but Nature's and ours? To bring about a humanly creative order, much that is now sacred must be destroyed. A great deal (hundreds of trillions' worth) of what now clutters up our earth and is regarded as most precious and everlasting must fall into disuse, become redundant or be transformed – to make room for new people-serving and earth-friendly technologies, relaxed, creative societies and democratically controlled work organization. There will have to be a new life-style. It is from the praxis of struggle to rediscover our lost humanness that we can get into perspective God's promise of a new heaven and a new earth.

Notes

1. D. Bohm and F. D. Peat, *Science, Order and Creativity*, New York 1987, 12.
2. See University Teachers for Human Rights (Jaffna), *Someone Else's War*, Colombo 1994; Rohini Hensman, *Journey Without a Destination*, Colombo 1994.
3. A translation of this into English is included in R. Panikkar, *The Vedic Experience*, Pondicherry 1883, 123–30.
4. E. O. Wilson, *The Diversity of Life*, Harmondsworth 1992, speaks of '*biophilia*, the connections that human beings subconsciously seek with the rest of life. To biophilia can be added the idea of wilderness, all the land and communities of plants and animals still unsullied by human occupation . . . ' (334). Cf. the 'holy ground' of the Moses story in Exod. 3.1–6.
5. See the discussion in Bohm and Peat, *Science, Order and Creativity*, 205ff., and Chapter 6 on creativity. According to Bohm, the general attitude referred to in n. 1 above 'leads us to focus always on particular problems, even when they are significantly related to a wider context'.

6. To grasp the significance for us today of Marx's understanding of nature and its relation to humanity requires studying in context the frequent references which occur often as asides in *Economic and Philosophical Manuscripts*, *Grundrisse*, and above all in his main critique of capitalism in *Capital*, Vol. 1. See K. Marx, *Grundrisse*, 323, 452, 461, 462, 705–706; Marx-Engels, *Collected Works*, Vol. 3, London 1975, 275ff., 336ff.; *Capital* Vol. 1, 43, 169, 177–184, 202, 206*, 333, 372, 441, 482n., 507, 508, 512–515, 603, 609.

7. Without getting 'back' to living in daily contact with the rest of Earthlife, city-dwellers will find it difficult to be certain about their origins, identity and responsibilities. R. Sheldrake, *The Rebirth of Nature*, London 1990, has some interesting things to say in this context.

8. When Wilson, *Diversity of Life*, writes that 'the race is on to develop methods to draw more income from the wildlands without killing them, and so to give the invisible hand of free-market economics a green thumb' (271), and gets to writing about 'bioeconomic analysis' (305ff.), his 'enduring environmental ethic' tends to be much too naive and uncritical.

9. See Ashis, Nandy (ed.), *Science, Hegemony and Violence: A Requiem for Modernity*, Delhi 1988.

10. In an unpublished paper, 'Aspects of Peasant Subsistence in India in the International Context of Commoditization of Production', Professor Utsa Patnaik of Nehru University has shown that China and India, while industrializing, have been expanding their domestic agricultural output over the past four decades, as well as per capita food output, at *much more* impressive rates than smaller Britain and Japan during industrialization.

11. A fuller discussion would have to take account of Dr Samir Amin's proposals for 'delinking'.

12. This is how to read Isa. 32.15–20 and Ezek. 34.23–31.

The Theological Debate on Ecology

Rosino Gibellini

'Creation on the brink of the abyss' – this was the title given by Günther Altner to one of the first theological studies of the ecological crisis,[1] to which the report of the Club of Rome, *The Limits of Growth* (1972), drew international attention.[2] It is interesting to note the change in perspective that the ecological problem has brought to the confrontation between Christianity and the modern world.

Whereas during the debate on secularization during the 1950s and 1960s theology defended the process of the formation of the modern world as a legitimate consequence of Christian faith, in the ecological debate begun in the 1970s which is now in full swing, theology is committed to relieving Christianity of responsibility for forming the anthropocentrism of the modern world. However, this is not so much an upheaval in the fronts as a more complex way of locating the relationship between Christianity and the modern world.

I Towards a theological ethic of responsibility

When developing his theological theory of secularization, in 'The Calamity and Hope of the Modern Age' (1953), Friedrich Gogarten time and again introduced the distinction between two modes of secularization: secularization as a legitimate consequence of the Christian faith, and secularism as a degeneration of secularization. In the first instance human freedom is 'freedom in chains' and thus is responsible for the world before God; in the second case, human freedom is freedom without chains. Autonomy becomes self-affirmation, as happens in the secularist anthropology of modernity.[3]

Gogarten did not specifically discuss the responsibility of human beings for the world and nature, which today goes under the name of the ecological crisis, but his formulation of the theory of secularization in the context of historical and systematic theology holds autonomy and responsibility together. In his *Anthropology in Theological Context* (1983), Wolfhart Pannenberg comments on Gogarten's thesis as follows: 'In delineating this exposition Gogarten did not have the ecological crisis in mind. But the way in which he distinguishes between secularization and secularism is illustrated in one of the most penetrating ways by precisely this crisis.'[4] Thus biblical Christian anthropology, if practised correctly, is better able to confront the ecological crisis than the absolute autonomy of modern culture.[5]

This observation, made in the context of systematic theology, is also confirmed in recent historical studies. In 'The Environmental Crisis – a Consequence of Christianity?' (1979), Ugo Krolzik has reconstructed with much learning the history of the interpretation of Gen. 1.28, 'subdue the earth', showing the complexity of the tradition which begins from the political mandate of dominion over the earth and identifying those extraneous elements which have been inserted in the modern era. In this interpretation particular mention should be made of Hugh of St Victor, who in his *Didascalion* (c. 1130) interpreted the new reality of the mechanical arts and spoke of man as the master and owner (*dominus et possessor*) who governs the visible world as though it were a machine. But in mediaeval thought dominion is subtended by a religious attitude in which human beings need illumination and divine revelation. From Bacon and Descartes onwards we find the beginnings of an inversion of the biblical idea of dominion in the form of a radical anthropocentrism, and it is this inversion, in connection with other cultural and technological factors, which ended up in the exploitation of nature. What is surprising is the great silence of theology and Christian ethics on the progressive and illegitimate expansionism of technological and industrial developments in the modern age.[6]

II The end of the anthropocentric conception?

However, theological reflection has proceeded beyond the secularization thesis, raising the question whether in the biblical conception human beings are the crown of creation. According to Jürgen Moltmann in his evocative work *God in Creation*, in the face of the ecological crisis some corrections also need to be made in the theological field.[7] In the Christian sphere, in fact a division of competences has largely developed: nature has been abandoned to science and technology, and theology has limited itself to history, which it interprets as a history of salvation. But by so doing, theology has failed to make faith culturally operative in creation. Hence the need to 'put a damper' on history, i.e. to move from the anthropocentrism of modern culture to the cosmological biblical theocentricity. Human beings and their history are inserted into the wider 'eco-system earth'. In his treatment, Moltmann follows the lines of an ecological doctrine of creation.

God has created the world with love and for his glory. Hitherto the doctrine of creation has been elaborated as a 'doctrine of six days' (the so-called 'Hexameron'), and no account has been taken of the seventh day. There is a need to recover the 'theology of the sabbath' and integrate it into the theology of creation. The 'sabbath doctrine' of creation includes the 'six days' work', but it also includes the seventh day, on which God rested and enjoyed his creation. However, in that case, it is not human beings who are the crown of creation; the sabbath is. According to Moltmann, it is the anthropocentric conception of the modern world which presents human beings as the pinnacle and crown of creation, and not the theocentric conception of the world of the Bible. In the creation story human beings are certainly the last creatures, but they are the last creatures before the sabbath and for the sabbath. Human beings are not the meaning of the world; the meaning of the world and of human beings lies in God: in God's glory.

However, for the moral theologian Alfons Auer, the author of an 'Ethics of the Environment' (1984, [2]1985, with an interesting new preface), theology needs, rather, to maintain its 'option for anthropocentrism'; indeed, 'nature realizes itself only in human beings and only in them does it reach its full significance'. By this transcendence over nature human beings not only represent themselves, but also bear responsibility for all living beings. If the 'radical anthropocentrism' inaugurated by Descartes subordinates nature to human beings and generates an anthropocentric arrogance and an imperialistic conception of nature, an 'anthropocentrism properly understood' constitutes the norm of all human behaviour, including that towards nature. And it is on this basis that Auer develops an ethic of the environment in the service of a concrete ecological ethic.[8]

The two studies just mentioned – Moltmann's systematic theology and Auer's essay in theological ethics – manifest two different approaches to the ecological question within Christian theology: on the one hand cosmological theocentricity (Moltmann), and on the other Christian anthropocentrism or creationist humanism (Auer). Is there an irreconcilable contrast here? There would be oppositions if the contrast were between a cosmocentric view and an anthropocentric view.

If one rigorously adopts the anthropocentric conception of modernity and the new philosophical-scientific cosmocentric, physiocentric or biocentric conceptions (to use some terms listed by Teusch in the *Lexikon der Umweltethik*, 1985), there is inevitably a conflict between the two conceptions, in that the first puts in the centre the totality of the cosmos, or nature, or life. But Christian biblical faith as faith in creation overcomes the alternatives of the anthropocentric conception, which asserts the privacy of human beings, and the cosmo-, physio-, biocentric conceptions which affirm the primacy of nature. The two lines of reflection indicated above – namely the line of cosmological theocentricity represented by Moltmann and the line of Christian anthropocentrism and creationist humanism represented by Auer – are not in conflict, even if they are distinct, in that both begin from the theological presupposition of nature as the creation of God, of *physis* as *ktisis* in the threefold dimension of protological creation, continuous creation and eschatological creation.[9]

Christian Link, the author of 'The World as Parable' (1975), indicates the cultural value of the thesis of creation with a brilliant quotation from Paul Cézanne: 'Nature is not on the surface but in the depths; the colours are the representation of this depth on the surface; they spring from the roots of the cosmos.'[10] This is as if to say that photography is always photography of the surface, but it is the picture as the art of colours which discerns the depths of the world. Similarly, one could say that theology, in interpreting the world as 'God's creation', reveals its depth, setting it against the horizon of the glory and the gift of God.[11]

III The conflict between nature and technology

Ecological theology has an acute sense of the situation of conflict which has developed between nature and technology, between creation and the technological achievements of humankind. How is it possible to settle this conflict?

According to the well-known theory of the Norwegian scholar Johan Galtung, two types of conflict can be distinguished: symmetrical conflicts and asymmetrical conflicts. Conflicts are symmetrical when the two entities in conflict are equivalent; they are asymmetrical when the two conflicting factors are not equivalent. Now in order to settle symmetrical conflicts it is important to use associative strategies. But to settle asymmetrical conflicts, as a first phase it is necessary to use dissociative strategies; associative strategies are to be used only in a second phase. One example of this is the conflict between employers and employees. In the past century the conflict was asymmetrical, and could not be settled until the workers had organized into trade unions; only at this point – when the conflict had become symmetrical – was it possible to bring associate strategies, i.e. negotiations, into play. In the context of this example the strike functions as a dissociative strategy aimed at arriving at a symmetrical conflict, whereas trade-union negotiations, which are employed at a second stage, aim directly at the settlement of the conflict.

On the basis of this theory, in his 'Ecological Theology' (1979), Gerhard Liedke shows how the conflict between creation and human beings is a conflict which is in many respects asymmetrical. In it the dominating entity is human technology and the threatened entity is creation. On the basis of this fact, which goes under the name of the ecological crisis, it would be premature to speak of peace with nature and harmony with the creation. The most urgent task to undertake is that of working, in a first phase, towards making the conflict symmetrical, encouraging the growth of the dimension which is threatened, in this case nature threatened by technology, going on in a second phase to a more harmonious settlement of the conflict: 'The emphasis of an ethic of creation in the ecological crisis should be . . . put at this point: the renunciation of force on the part of human beings; a reduction in the excessive pressure of violence to which we subject the non-human creation. The aim of the renunciation of force is to arrive at a more symmetrical configuration of the ecological conflict. Certainly we cannot lose sight of the final goal of arriving at the outline of a harmony with nature, but this belongs to a second phase of the settlement of the conflict, and this second phase is not yet in sight'.[12]

Whereas Liedke spoke of an ethics of solidarity in the conflict between creation and human technology, in his vast study on 'Creation', Christian Link speaks of an ethic of self-limitation. To be a creature means to exist within determined limits which are not ours to set; they have been laid down by the Creator to creatures, so that within the sphere of what has become technically possible there is a need to make a distinction between what is ethically practicable and what is not. As Link puts it: 'Today gene technology and the industrial use of nuclear energy are obvious problematical areas in which we are called to decide on these limits. Therefore in both cases the power of control that we claim has taken on a qualitatively new dimension . . .'[13]

We are a long way from the theological interpretations of the biblical mandate to dominate the earth which have been offered from the beginning of the modern age to the theology of secularization, and of which this quotation from Schleiermacher (1799) should be adequate illustration. 'We hope from the perfecting of the sciences and the arts . . . that they will transform the physical world, and that all this can be governed from the spiritual world, in an enchanted palace from which the God of the earth will need only to pronounce a magical formula and to push a button for what he ordained to be realized.'[14] With theology in an ecological perspective there is a move from the category of control, in which nature is completely at the disposal of human beings as material which can be manipulated and be the object of technology, to the category of the conservation of creation and of co-operation with it.

IV Ecofeminism

According to Anne Primavesi in her *From Apocalypse to Genesis* (1991), ecofeminism is a matter of bringing into play an ecological paradigm from a feminist perspective.[15] From the beginning, feminist theology has argued that a redefinition of the relationship between man and woman in terms of reciprocity (rather than hierarchy) also leads to a redefinition of the relationship between human beings and nature. In her *Sexism and God-Talk* (1983), Rosemary Ruether wrote: 'The "brotherhood of man" needs to be widened to embrace not only women but also the whole community of life.'[16] The superior forms of life must become aware of their dependence on ecological harmony and their responsibility towards the ecological community. Therefore feminist theology sets out to rethink the whole Western tradition of the hierarchical chain of being which is transformed into a hierarchical chain of command from the superior to the inferior; it is within this wider hierarchical scheme that the domination of the female by the male is set; thus feminist theology makes itself ecological theology: 'This theology must question the hierarchy of human over nonhuman nature as a relationship of ontological and moral value. It must challenge the right of the human to treat the nonhuman as private property and material wealth to be exploited.'[17]

This is a theme which Rosemary Ruether has taken up again in her study *Gaia and God. An Ecofeminist Theology of Earth Healing* (1992), in which she identifies the ecological line in the Bible as a way of healing the earth, and dissociating it from the culture of domination. For this Chicago theologian, two lines of biblical thought and two Christian traditions offer resources for use by an ecological spirituality and practice – these can no longer be differentiated. In particular there are the traditions of the covenant: the God who becomes present in the historical acts of liberation is at the same time the God who 'has made heaven and earth', in such a way that human beings are inserted into the 'covenant of creation'; and the sacramental tradition, for which the God of creation is a relational God, though a God who is to be interpreted as the 'creative matrix of all things', from which we proceed and to which we go.[18]

Ruether re-emphasizes her thesis in a recent article in *Ecotheology:*

Voices from South and North (1994), arguing that to create an ecological culture and society we must transform relations of dominion and exploitation into relations of reciprocal support. This transformation will not take place without a corresponding change in our image of God, in our image of the relationship between God and creation in all its dimensions.[19]

Another important study, by the North American theologian Sally McFague, *The Body of God. An Ecological Theology*, also follows this line. In it she proposes a redefinition of God and God's relationship with the world in the light of the changed view of the world introduced by postmodern science. If postmodern science has brought about a move from the mechanistic cosmological model to the organic cosmological model, a theology of nature understood as ecological theology is called to see the transcendence of God embodied, and thus to conceive of the world as God's body: 'In this one image of the world as God's body, we are invited to see the creator *in* the creation, the source of all existence in and through what is bodied forth from that source.'[20] This brings about a radicalization of the incarnation: God is not present only in Christ, even if this presence is a paradigmatic one, but is present in the totality of the world as God's body. If the universe in its totality is the body of God, God in the divine immanent transcendence is the spirit of the life of the world, as God's body. This would not be a pantheistic conception but, according to the author, a panentheistic one: 'Everything that is is *in* God and God is *in* all things and yet God is not identical with the universe, for the universe is dependent on God in a way that God is not dependent on the universe.'[21]

Both Ruether's and McFague's approaches in substance present a radicalization of the *immanent* transcendence of God in the world, which is capable of generating a new vision of the world of a kind that provokes an awareness of the inter-relatedness of all things and the ecological responsibility which follows from this.

However, Elisabeth Green, while recognizing the theological and cultural contribution of ecofeminism, has a criticism. Whereas Christianity has always been very careful to distinguish between God and the world, thus safeguarding their distinct ontological realities, it is not clear how such expressions of ecofeminism, despite their affirmations of panentheism, really avoid pantheism, the identification of nature with God, which in the end reduces both God and the world.[22] For Elisabeth Moltmann-Wendel, too, there is a need to avoid expressions like 'body of God' with reference to the world because they are not biblical; what is important, though, is the task of recovering the biblical and theological depth associated with the earth[23] (and with it of the body[24] and the senses[25]), overcoming the reductionism to which the concept of earth has been subjected, in which it has been associated with dust and fallenness, thus encouraging its manipulation and exploitation.[26]

V Ecology as a theme of liberation theology

Though feminist theology has taken up the topic of ecology as one of its major themes right from the beginning, the topic of ecology is almost completely absent even from the vast work *Mysterium Liberationis* (1990), which presents the basic concepts of liberation theology in thematic monographs. It emerges after the Rio de Janeiro conference on ecology in 1992 and the articles collected in Leonardo Boff's 'Ecology, World and Mysticism. The Emergence of a New Paradigm' (1993). Here liberation theology, too, has decisively taken up the ecological theme. It has done so in its own way, specifically from the perspective of the Third World. The ecological crisis confirms liberation theology in its analysis of the crisis at the depths of our system of life, our model of society and development. Boff argues that there is a need to welcome the problem which has arisen from the awareness of the rich, to present it in a different way and to provide a different solution in the interest of all human beings and nature, beginning from those who are most threatened among human beings and among the other beings of creation.[27] The recovery of the theology of creation must go beyond environmentalism, which fights for the defence of the environment, and beyond conservationism, which fights for the conservation of living species: ecology in the perspective of liberation questions the model of development. It argues that the industrialized countries, almost all in the northern hemisphere, are responsible for 80% of the pollution of the earth (the USA alone for 23%);[28] it has put forward a plan for an ecologico-social democracy in which social justice must include ecological justice, as is evident in the case of Amazonia, where ecological exploitation has also become social exploitation.

Theology and the Christian church today are called on to arouse themselves from what could be defined as 'forgetfulness of creation' and to develop all the theological and ecclesial potential of creation faith, which is also cultural and political.

Translated by John Bowden

Notes

1. G. Altner, *Schöpfung am Abgrund. Die Theologie vor der Umweltfrage*, Neukirchen 1974. Cf. id., 'Technisch-wissenschaftliche Welt und Schöpfung', in

Christliche Glaube im moderner Gesellschaft, Enzyklopädische Bibliothek XX, Freiburg, Basel and Vienna 1982, 85–118 (with bibliography).

2. First contributions in the theological field include J. Cobb, *Is It Too Late? A Theology of Ecology*, Beverly Hills, Ca. 1972; T. S. Derr, *Ecology and Human Liberation. A Theological Critique of the Use and Abuse of our Birthright*, Geneva 1973.

3. F. Gogarten, *Verhängnis und Hoffnung der Neuzeit. Die Sekularisierung als theologisches Problem*, Stuttgart 1953 (esp. ch. 3).

4. W. Pannenberg, *Anthropology in Theological Context*, Edinburgh 1985, 76.

5. There are some authors who blame Christianity for the ecological crisis; they include L. White, *The Historical Roots of Our Ecological Crisis*, New York 1970; C. Amery, *Das Ende der Vorsehung. Die gnadenlosen Folgen des Christentums*, Reinbek 1972; E. Drewermann, *Der tödliche Fortschritt. Von der Zerstörung der Erde und des Menschen im Erbe des Christentums*, Regensburg 1981.

6. J. Krolzik, *Umweltkrise – Folge des Christentums?*, Stuttgart 1979.

7. J. Moltmann, *God in Creation. An Ecological Doctrine of Creation*, London and New York 1985. Cf. also id., 'Die Erde und die Menschen. Zum Theologischen Verständnis der Gaja Hypothese', *Evangelische Theologie* 53, 1993/5, 420–38.

8. A. Auer, *Umweltethik. Ein theologischer Beitrag zur ökologischen Diskussion*, Düsseldorf 1974, [2]1975; cf. especially the preface to the second edition, in which he comments on the discussion.

9. For the discussion cf. 'Ist der Mensch die Krone der Schöpfung? Drei Theologen beziehen Position: A. Auer, E. Drewermann und J. Moltmann', *Publik-Forum*, special issue on the environment, 31 May 1985; J. L. Ruiz de la Peña, *Teologia de la creación*, Santander 1986; Associazione Teologica Italiana, *La Creazione. Oltre l'antropocentrismo?*, Padua 1993; D. Hervieu-Léger (ed.), *Religion et Ecologie*, Paris 1993; R. Gibellini, 'La svolta teologica nella percezione del cosmo', *Rocca* (Assisi), 1 August 1994, 49–52.

10. Cf. C. Link, 'Die Transparenz der Natur für das Geheimnis der Schöpfung', in G. Altner (ed.), *Ökologische Theologie. Perspektiven zur Orientierung*, Stuttgart 1989, 166–95: 177 (this is an important work for the variety of its perspectives on ecological theology).

11. Works which present an ecological spirituality include M. Fox, *Original Blessing*, Santa Fe, NM 1983; S. McDonagh, *To Care for the Earth. A Call for a New Theology*, London 1986; T. Berry, *The Dream of the Earth*, San Francisco 1988; I. Brandly, *God is Green. Christianity and the Environment*, London 1990; H. Kerssler, *Das Stöhnen der Natur. Plädoyer für eine Schöpfungsspiritualität und Schöpfungsethik*, Düsseldorf 1990.

12. G. Liedke, *Im Bauch des Fisches. Ökologische Theologie*, Stuttgart 1969, 165–78; id., 'Schöpfungsethik im Konflikt zwischen sozialer und ökologischer Verpflichtung', in *Ökologische Theologie. Perspektiven zur Orientierung*, 300–21: 313.

13. C. Link, *Schöpfung. Schöpfungstheologie angesichts der Herausforderungen des 20. Jahrhundert*, Gütersloh 1991, 487. Cf. also W. Korff, *Kernenergie und Moraltheologie*, Frankfurt 1979; P. Spescha, *Energie, Umwelt, Gesellschaft: Gerechtigkeit und Frieden*, Fribourg 1983.

14. Quoted in Liedke, *Im Bauch des Fisches* (n. 12), 69.

15. A. Primavesi, *From Apocalypse to Genesis. Ecology, Feminism and Christianity*, Tunbridge Wells 1991, 36.

16. R. Radford Ruether, *Sexism and God-Talk. Toward a Feminist Theology*, Boston and London 1983, 87. Cf. also ead., *New Woman. New Earth. Sexist Ideologies*

and Human Liberation, New York 1975.

17. Ibid., 85.

18. R. Radford Ruether, *Gaia and God. An Ecofeminist Theology of Earth Healing*, San Francisco and London 1992, esp. 205–53.

19. R. Radford Ruether, 'Eco-feminism and Theology', in D. G. Hallman (ed.), *Ecotheology. Voices from South and North*, Geneva 1994, 204.

20. S. McFague, *The Body of God. An Ecological Theology*, Minneapolis and London 1993, 133–4.

21. Ibid., 149.

22. E. Green, 'Ecofeminism and Theology', *Jahrbuch der Europäischen Gesellschaft für theologische Forschung von Frauen*, Mainz and Kampen 1994/2, 53f.

23. E. Moltmann-Wendel, 'Rückkehr zur Erde', *Evangelische Theologie* 53, 1993/5, 406–20.

24. E. Moltmann-Wendel, *I am my Body. New Ways of Embodiment*, London and New York 1994.

25. H. Pissarek-Hudelist and L. Schottroff (eds.), *Mit allen Sinne glauben*, Gutersloh 1991.

26. Cf. also D. Sölle, *To Work and to Love. A Theology of Creation*, Minneapolis 1984; C. Halkes, *Das Anlitz der Erde erneuern. Mensch, Kultur, Schöpfung*, Gütersloh 1990.

27. L. Boff, *Ecologia – Mundialidade – Mistica. A emergéncia de um novo pardigma*, Petropolis 1993. Cf. also E. Dussel, *Etica comunitaria*, Petropolis 1986 (esp. ch. 12).

28. Boff, *Ecologia* (n. 27), 21.

Special Column 1995/5

1. Has the Papacy an Ecumenical Future?

Jürgen Moltmann

The papal encyclical on ecumenicity, *Ut unum sint*, in essence offers only the well-known Roman view of the 'Petrine service'. But the fact that this is published in a special encyclical on ecumenicity can only be understood by non-Roman Catholic readers to mean that here the papacy is being made a topic for discussion with other churches. So we return to the old question whether the ecumene has a papal future. 'We are perfectly aware that the Pope is the greatest obstacle on the way to ecumenism,' declared the honest, self-critical Paul VI in 1967. That is not John Paul II's starting point. A Catholic-Lutheran Dialogue Commission in the USA summed up its findings in the thesis 'that the papal primacy, renewed in the light of the gospel, need not be an obstacle to reconciliation'. If the papacy is the subject of ecumenical discussion, constructive ideas must be presented about a 'papacy renewed in the light of the gospel' and we cannot limit ourselves to more or less friendly but trivial reactions to this Roman self-portrait.[1]

1. The way in which the encylical begins with the 'anti-Christian tendencies' of secular modernity is one-sided, inappropriate and characteristic of East European Catholicism. However, the remarks about the Christian martyrs of our century whom we all revere and the theology of the cross to which all of them bear witness are good and ecumenical. And in mentioning the Orthodox, Protestant and Catholic martyrs in the European resistance by name, we should not forget the Catholic martyrs in the Catholic dictatorships of Latin America, from whose number Archbishop Oscar Arnulfo Romero might be mentioned as a representative. The poor

of the crucified Christ embrace us and invite all who are weary and heavy laden to share the one eucharist with them. Ecumenism is at heart sharing the meal at the Lord's table; everything else can be discussed after that.[2] Who has the right to hold back these outstretched arms of Christ and exclude from the Lord's table Christians who hear the call?

2. Of course the Pope offers his 'Petrine service' as his commitment to ecumenism. The great non-Roman churches are also prepared for a *communio cum Petro*, the Bishop of Rome. But they are not prepared for a *communio sub Petro*. There will be no retrograde ecumenism (*unitatis redintegratio*) leading to submission to the universal episcopate of the Pope and his authority in matters of faith and morals. In the two-thousand-year history of the church, the Popes have made no contribution worth mentioning to the ecumenical unity of Christianity. The excommunications of the great Orthodox, Protestant, Anglican and Old Catholic churches came from Rome. The non-Roman churches have not forgotten those of their number who were martyred by Rome. Nor has the present Pope made any particular contribution to the ecumenical unity of the church: when after the collapse of the Soviet empire there was a unique occasion for a pan-Christian council in Europe, instead he proclaimed a Roman Catholic 're-evangelization of Europe' with a special focus on the Orthodox peoples and the churches in Eastern Europe which were once united with Rome. He let slip the moment when the Petrine service could unselfishly have been offered to European Christianity. That was a deep disappointment for many Christians in Europe, Catholic and non-Catholic.

3. Can a Petrine service to the *una sancta ecclesia* seriously be exercised by a head of state? When a Catholic colleague in Tübingen called on the head of his church to lay down his office he was admonished through state channels that it was wrong for him as a German state official to call on a foreign head of state to lay down his office. The twofold representation of the Pope by conferences of bishops and the nuncio, i.e. by church politics and Vatican diplomacy, is far removed from what Christ ever said about Peter. It is a relic of the mediaeval idea of the church state which corresponds neither to the nature of the church nor to the spirit of modern times. Vatican diplomacy did not bring abut any peace with the Orthodox in Bosnia, and at the UN world population conference it prevented statements in favour of women. Does the papacy have an ecumenical future as long as the Pope is at the same time a head of state?

4. Not least, the encyclical raises the question whether the papacy has Catholic future. The centralistic system of authority which works both ecclesiastically and politically is not appropriate to the wonderful, world-

wide, trans-national, inter-cultural Catholic community which transcends the races, as is shown by the forced appointments of Opus Dei people as bishops and the measures taken against loyal Catholic theologians. The unity of the Catholic community is much wider and more powerful than the centralistic unity of the papacy. That leads to the critical question in which we all have an interest: is the Vatican papal system good for the Catholic Church?

Notes

1. The *concrete* perception of the Roman papacy is the essential obstacle on the way to a credible witness to the unity of Christianity in world society.
2. Canon 844 of the 1983 Catholic Code of Canon Law limits 'ecumenical hospitality' quite considerably.

Ut unum sint. Remarks on the New Papal Encyclical from an Orthodox Perspective

John Panagopoulos

In its basic concerns, the new papal encyclical hardly comes as a surprise to the Orthodox Church. Indeed, for the most part it is a supplementary commentary on and a consistent development of the Second Vatican Council's decree on ecumenism, *Unitatis redintegratio*. However, there is no mistaking the fact that the encyclical is strongly marked by the spirit of openness, responsibility, living hope, humility and self-accusation (34). The understanding of church unity as a gift of Jesus Christ through his Spirit (35), the elucidation of historical thought which is called for (2), the betrayal of Catholic believers (11), the need for inner conversion and renewal (15), the mutual enrichment, the prayer, the holiness, the love, the truth, the martyrdom, brotherly dialogue as conditions of unity and the Pope's call to all Christians for forgiveness (34, 88), in other words the priority of 'spiritual ecumenism' (21), all this and more are authentic and encouraging Christian beginnings which must be taken seriously. The first part of the encyclical in particular, which can be seen as a 'spiritual introduction' to ecumenical matters, is a welcome ecumenical contribution.

The encyclical accords a relatively large amount of space to the Orthodox Church (50–61). Whereas it is conceded that the other Christian communities have preserved important elements of Christian truth (10–13), the Orthodox Church is regarded as a sister church, the other 'lung' of the body of Christ (54), but separated from the Roman Catholic Church. Its apostolic succession and its sacraments are unconditionally recognized, while its spiritual and liturgical riches are appropriated with gratitude (50, 57). However, despite these concessions there is no mistaking the fact that the Orthodox Church is not thought to show the fullness of truth, any more than the Christian communities which emerged from the Reformation, as long as they do not enter into communion with the Roman see. That throws out the baby with the bath water. The Roman Catholic Church emerges as the judge and final authority when declaring whether Christian communities are the church.

The real surprise in the encyclical is certainly its uncompromising endorsement of the Vatican II Decree on Ecumenism. Its main theme is that 'the communion of all individual churches with the church of Rome is the necessary presupposition of unity' (97). The primacy of the Bishop of Rome is grounded in God's saving plan (ibid., 92, etc.), and is understood as the safeguard of church unity, the tradition of faith, sacramental and liturgical actions, mission, church order and Christian life generally (94). Only communion with the successors of Peter guarantees the fullness of the *una sancta*. So any discussion of church unity must begin from the authority and indisputable clarity of the Petrine office, which God has founded as a 'permanent, visible and fundamental principle of unity' (88, cf. 95). The concession that important elements of Christian truth and sanctification are to be found in other churches (but who defines them?) tends to create confusion, since they are seen only as partial truths bestowed from the fullness of the Roman Catholic Church, truths which can be fully effective only in communion with it (14). So at best this has to be taken seriously as a diplomatic gesture. As usual, the limits of the church are laid down narrowly on the lines of Roman Catholic canon law. This exclusive understanding of the church is not in principle ecumenical.

Unfortunately Orthodox Christians can only indicate their boundless disappointment at this encyclical. For this traditional understanding of the Roman Catholic Church and of unity has been a stumbling block since as long ago as the fifth century, and despite the intensive theological conversations we have not got one step further. Nor does it seem that we shall be able to do so. In this sense the encyclical sets a new insuperable obstacle in the way of ecumenism. The *jure divino* universal claim of the Roman Catholic Church, grounded on the postulate of a Petrine or papal primacy, has already been settled as far as the Orthodox are concerned, in terms of exegesis, dogma and church history. If the Roman Catholic Church would be prepared to understand the primacy in the sense of *primus inter pares*, as was the case in the church of the first eight centuries, and not as a dogmatic question of faith, the way would certainly be open to communion, especially as the Orthodox cannot give up the notion of ecclesial primacy. In the sense of a primacy of honour, the supreme magisterium would be attributed to the synodical structure of the church and not to a historical church, i.e. the See of Rome. To judge from the encyclical, the Roman Catholic Church is evidently not prepared to take such a truly ecumenical step. For all their respect for the person of the Pope, to the Orthodox it is an 'ecumenical sin' heedlessly to present the papal primacy as the foundation of and condition for church unity. If the Roman Catholic Church is going to insist on this stubbornly, then it must expect the inevitable verdict that it does not envisage church unity from the perspective of the *una sancta*, but from that of a confessional church. As Orthodox understand it, papal primacy can only be accepted as a theologumenon of the Latin church, not as an essential principle of the *una sancta*.

The Orthodox hardly expect the Roman Catholic Church to change its spots. It is obligated to a heritage which it cannot deal with lightly. Orthodox fully understand this. However, they did expect that their 'sister' Roman Catholic Church would have taken seriously the claim of Orthodoxy to manifest the *una sancta* in its life, thought and ethos and in its liturgy faithfully and in unbroken continuity, and to have drawn the necessary consequences from that. This would have made the 'sisterhood' of the two churches more than rhetorical diplomacy. In the end it is a matter of looking at facts, of establishing objectively which church, the Orthodox Church or the Roman Catholic Church, has preserved the fullness of Christian truth faithfully and unbrokenly to the present day. It is certainly true that the Orthodox Church and the Roman Catholic Church are agreed in principle on the great questions of faith, apart from the *filioque*, papal primacy and papal infallibility. However, one must ask seriously on what this accord is based, if not on the Christian heritage shaped by the Greek church fathers above all of the fourth and fifth centuries, which was taken over by the Latin church.

The new encyclical makes the Orthodox world confused about the future of Christian unity. Its 'spiritual ecumenism' certainly does not remove the justified suspicion of the Orthodox. Certainly they are firmly resolved to continue dialogue with the Roman Catholic Church, as the Ecumenical Patriarch Bartholomaios again confirmed on his last visit to the Vatican; however, since then this dialogue has lost any real point. In any case the climate has been made more difficult by the unfortunate interventions of the Roman See in Eastern Orthodox countries after the changes in Eastern Europe. The tricky question of the Uniate churches is also left open in the encyclical (60), despite the joint condemnation of Uniatism in the Freising Document. Nevertheless, the Orthodox do not want to doubt the seriousness of this encyclical, although the confessionalist Roman Catholic model of unity promises nothing. Church unity certainly cannot be brought about through formal attachment to a historical church; it must be communion (*koinonia*) in the whole Christian truth, as it has been believed and experienced always, by all people, everywhere. For the Orthodox there is still just one demand which emerges from the living potential of the *una sancta*: an inclusive eucharistic charismatic understanding of the church and a eucharistic, synodical unity grounded on the one undivided church of the first eight centuries. So it should be accepted that, rather than the papal claim (97), the communion of all local churches with the faith and life of the church of the first eight centuries is the necessary presupposition for unity. This model has become much more topical after the encyclical and it is now the urgent task of the Orthodox to make their ecumenical contribution with prayer (as they always do in their liturgy), with spiritual watchfulness and humility, with dialogue in love and truth (Eph. 4.15). This eucharistic-synodical principle of unity presupposes that church unity governs a spiritual growth towards Christ with the power of the Holy Spirit, in other words a departure (*exodos*) from one's own historically conditioned confessionality (particularity), a common way (*synodos*) of change (II Cor. 3.18), and entry (*eisodos*) as eucharistic *koinonia* into the sanctuary of the kingdom of God in which God will be all in all (1 Cor. 15.28).

The editors of the Special Column are Miklós Tomka and Willem Beuken. The content of the Special Column does not necessarily reflect the views of the Editorial Board of Concilium.

Contributors

JULIO DE SANTA ANA was born in Montevideo (Uruguay) in 1934. He studied law and theology and graduated in the Buenos Aires Faculty of Theology in 1956. Later he obtained a doctorate in Sciences of Religion from the University of Strasbourg. He was the general secretary of the Movement 'Church and Society in Latin America' (ISAL). He then worked as secretary of studies for the Commission for the Churches' Participation in Development (CCPD) at the World Council of Churches, of which he was Director from 1979 to 1982. Between 1983 and 1993 he was Co-Director of the Ecumenical Centre for Evangelization and Education Services (CESEP) in São Paulo, Brazil, where he was also Professor of Social Sciences and Religion in the Methodist University of São Paulo, of which he is now an emeritus professor. He is the author of many articles and a number of books, translated into different languages: *Good News to the Poor*; *Towards the Church of the Poor*; *Ecumenismo y Liberación*, etc.

Address: 15 Chemin de l'Erse, CH-1218 Grand Saconnex, Geneva, Switzerland.

BERTA G. RIBEIRO was appointed associate professor at the National Museum, Federal University of Rio de Janeiro, in 1988. She also teaches anthropology of art at the Fine Arts school of the same university. She holds a degree in geography and history and a doctorate in social anthropology from the University of São Paulo. Her special subject is indigenous ethnology, with further specialization in indigenous art, handicrafts, technology and economy. Her recent studies of ecological anthropology have been carried out with the Desâna Indians of the Upper Rio Negro, in Amazonia. Her published works include: *Arte Plumária dos Indios Kaapor* (1958) (with Darcy Ribeiro); *Diário do Xingu*: *Dicionârio do Artesonato Indígena: Arte Indígena, Linguagem Visual/Indigenous Art: Visual Language*, bi-lingual ed.; plus some sixty articles in specialist journals. She has mounted several exhibitions in Brazil and Italy, of which the most important was *Amazonia Urgent: Five Centuries of History and*

Ecology, the book of which was also published in Portuguese and English.

Address: Rua Souza Lima 245, apto. 901, 22081–010 Copacabana, Rio de Janeiro, RJ, Brazil.

SYLVIA MARCOS is a Mexican women's rights activist who lives and teaches in Cuernavaca. She trained in clinical and social psychology and Latin American studies and had a post-doctoral fellowship at Harvard Divinity School; she has lectured widely in the United States, often as visiting professor. At present she is Director of the Centre for Psychoethnological Research, Cuernavaca, a research associate in medical anthropology at the Instituto Nacional de Antropología e Historia and in the Programa Interdisciplinario de Estudios de la Mujer, Colegio de Mexico, and practises privately as a clinical psychologist. Her books include *Manicomios y Prisiones* (1983), *Alternativas a La Psiquiatria* (1982), and *Antipsiquiatria y Politica* (1980). She is also international editor of Gender in Society Religion.

Address: Centro de Investigaciones Psicoentologicas, Apdo 698, CP 62000, Cuernavaca, Morelos, Mexico.

CHRISTOPH UEHLINGER was born in 1958, and after studying in Fribourg, Bern, Jerusalem and London has been a lecturer in Old Testament exegesis and the biblical world in the theological faculty of the University of Fribourg. He also lectures regularly in Haiti. His main theological interest is the relationship between culture, religion and social development in the Bible and the ancient Near East, with particular reference to iconographic and archaeological sources. His publications include *Weltreich und 'eine Rede'. Eine neue Deutung der sogenannten Turmbauerzählung (Gen. 11.1–9)*, Orbis Biblicus et Orientalis (1990); (with O. Keel and M. Küchler), *Orte und Landschaften der Bibel I, Geographisch-geschichtliche Landeskunde* (1984); (with O. Keel), *Göttinnen, Götter und Gottessymbole. Neue Erkenntnisse zur Religionsgeschichte Kanaans und Israels aufgrund bislang unerschlossener ikonographischer Quellen*, QD 134 ([2]1993).

Address: Biblical Institute, Université Miséricorde, CH 1700 Fribourg, Switzerland.

JULIA ESQUIVEL VELÁSQUEZ is a Guatemalan teacher. She studied pastoral theology in the Latin American Biblical Seminary and in the Bossey

Ecumenical Institute (Costa Rica and Switzerland). She has published three books of poetry.

Address: Avenida La Garita, Andador 22, conjunto 9–3, Villa Coapa, México DF 14390, México.

LEONARDO BOFF was born in Concórdia, Brazil, in 1938. He studied in Curitiba, Petrópolis and Munich. He taught systematic theology for twenty years in Petrópolis and now lectures in ethics at the University of Rio de Janeiro. His many works translated into English include *Jesus Christ Liberator* (1979) and *Church: Charism and Power* (1982), which earned him a year's imposed silence from the Vatican, during which he composed *Trinity and Society* (1988). *A New Evangelization: Good News to the Poor* (UK *Good News to the Poor*) appeared in 1992 and his latest work, *Ecology and Spirituality*, is published in 1995.

Address: Pr. Martins Leão 12/204, Alto de Boa Vista, 2051–350 Rio de Janeiro R.J., Brazil.

JOSÉ RAMOS REGIDOR was born in Spain in 1930, but has lived in Italy for more than forty years. He has taught theology in the Pontificia Universita Salesiana, Rome, and between 1973 and 1990 worked at the IDOC Centre in Rome. He is now working in the 'North-South Campaign: Biosphere, Survival and Debt' and is a member of the Christian base community of San Paolo, Rome. His publications include *Il Sacramento della Penitenza* (1971) and *Gesù e il Risveglio degli Oppressi. La Sfida della Teologia della Liberazione* (1981). In recent years he has been interested in questions related to the interaction between social questions, ecological questions and women's questions. With Alessandra Binel he has edited the volume *Dissenso sul mondo* (1992).

Address: Via degli Strengaré 25, 00186 Rome, Italy.

BASTIAAN WIELENGA was born in the Netherlands in 1936 and studied Reformed Theology in Kampen. Between 1961 and 1971 he worked in a Dutch ecumenical parish and at Hendrik Kraemer Haus, an ecumenical centre in Berlin. His doctoral thesis, 'Lenin's Way to Revolution. A Comparison with S. Bulgakov', under Helmut Gollwitzer, was published in 1971. Between 1972 and 1974 he was research fellow in the Christian Institute for the Study of Religion and Society in Bangalore, and since 1975 he has taught social analysis and biblical theology at Tamilnadu

Theological Seminary, Madurai, South India. He has written in the areas of Marxist theory, biblical theology, international developments and new social movements.

Address: Centre for Social Analysis, 37 Ponmeni Narayan Street, Somasundaran Colony, Madurai 625 016, Tamilnadu, India.

EDUARDO GUDYNAS was born in Montevideo in 1960. He is the Coordinator of Environment and Development in the Franciscan and Ecological Centre (CIPFE) and a researcher in the Latin American Centre for Social Ecology (CLAES), both in Montevideo, Uruguay. He is joint author of *La Praxis por la vida*, a manual of methodologies in social ecology, published both in Latin America and Spain. He is the promoter of the idea of a multiversity as an alternative centre of studies and practice in Montevideo, and was its first academic coordinator, from its foundation in 1989 until 1994.

Address: AyD–CIPFE, Canelones 1164, Montevideo, Uruguay.

CHARLES RICHARD HENSMAN was born in Nallur in Sri Lanka, and lives outside Colombo. He did a year of theological study at Yale Divinity School, 1956–7. He was Research Secretary of the Overseas Council of the Church of England, 1958–61. As an editor, lecturer, lay theologian, etc., and organizing workshops, he has worked on the assumption that it is mainly in how the laity live out in the real world, the social and economic order, their faith hope and love in Christ that the world can come to discern God at work. He is a member of the Jubilee Group and EATWOT, and Vice-President of the Sri Lankan Association of Theology (SLAT). Among his books are: *China: Yellow Peril? Red Hope?*, London 1968; *From Gandhi to Guevara*, London 1970; *Rich Against Poor*, London 1971; *Agenda for the Poor: A Reading of Luke*, Colombo 1992; *New Beginnings: The Ordering and Designing of the Realm of Freedom* (2 vols), Mt Lavinia 1992, 1994, and several on Sri Lanka.

Address: 23/1 Dharmapala Road, Mount Lavinia, Sri Lanka.

ROSINO GIBELLINI is editorial director of Editrice Queriniana, Brescia. He gained his doctorate in theology at the Pontifical Gregorian University in Rome and his doctorate in philosophy at the Catholic University, Milan. He has written books on Teilhard de Chardin, Jürgen Moltmann and Wolfhart Pannenberg and an important study of modern theology, *La*

Teologia del XX secolo (1992). His *The Liberation Theology Debate* (1987) has appeared in English, as have other books edited by him: *Frontiers of Theology in Latin America* (1980) and *Paths of African Theology* (1994).

Address: Editrice Queriniana, Via Piamarta 6, 25187 Brescia, Italy.

Members of the Advisory Committee for Theology of the Third World

Directors

Leonardo Boff OFM	Rio de Janeiro	Brazil
Virgil Elizondo	San Antonio, Texas	USA

Members

K. C. Abraham	Bangalore	India
José Miguez Bonino	Buenos Aires	Argentina
J. Russel Chandran	Bangalore	India
Frank Chikane	Braamfontein	South Africa
Zwinglio Mota Dias	Rio de Janeiro RJ	Brazil
Enrique Dussel	Mexico, DF	Mexico
Gustavo Gutiérrez	Lima	Peru
François Houtart	Louvain-la Neuve	Belgium
Beatriz Melano Couch	Buenos Aires	Argentina
Ronaldo Muñoz	Santiago	Chile
Alphonse Mgindu Mushete	Kinshasa, Limeta	Zaire
M. A. Oduyoye	Geneva	Switzerland
Juan Hernandez Pico SJ	Mexico, DF	Mexico
Aloysius Pieris SJ	Gonawala-Kelaniya	Sri Lanka
Pablo Richard	San José	Costa Rica
Anselme Titianma Sanon	Bobo-Dioulassa	Upper Volta

Members of the Board of Directors